Portrait of John Smeaton about 1759 attributed to Benjamin Wilson. © Royal Society

Published by Wakefield Historical Publications
(publication number 46), a body comprising
Wakefield Historical Society and Wakefield Council,
formed in 1977 to publish monographs of regional significance.

ISBN 978 0 901869 49 4

Wakefield Historical Publications
18 St John's Square, Wakefield WF1 2RA
www.wakefieldhistoricalsociety.org.uk/publications/wakefieldhistoricalpublications

Designed by Paul Buckley of Riasca
Printed by Leeds Graphic Press Ltd

Front cover:
1. Illustration from the Subscription of Japhet Lister to the Upper Calder Navigation, 1763.

CONTENTS

PREFACE

The Superintendent's Journal, transcribed in this book with additional illustrations, describes the day-to-day progress of work on the upper Calder Navigation. It was prepared by John Smeaton (1724-1792) for his employers, the Calder Navigation Commissioners, over the first four years of construction, 1760-1763. The navigation began at Wakefield and the work was based there for much of this time, so that the Journal provided the Halifax gentlemen, over twenty miles up the river, with up-to-date information. The weekly reports were copied by a clerk into a bound book of about 170 sides of script, with days of the week and dates noted in the margins. This Journal is available to researchers at the National Archives at Kew and on microfilm at Calderdale Archives, Halifax, as part of the Calder and Hebble Archive.

Smeaton's Journal provides an exceptional insight into the way this project developed, from the first entry at the end of May 1760 when he reports on his visit to the recently-established Wakefield yard, to the last entry in November 1763 when the navigation had reached beyond Brighouse. It describes the activity at each location as the cuts, weirs and locks were gradually constructed, the many challenges met, and how these were tackled. Smeaton's regular, detailed account allowed the commissioners to see how his time was occupied, what difficulties he and his workmen faced, the reasons for his decisions, and the progress of the project.

A long introduction using a number of additional primary sources supplements the Journal transcription. The opening chapters describe the various events, issues and personalities which influenced the story of this navigation from 1740/41, when the idea was seriously promoted but successfully opposed. Other records in the Calder and Hebble archive cover the following period: minutes, accounts, and letters as well as Parliamentary evidence, from the cautious renewal of the pursuit of an Act of Parliament in the 1750s to the first years of the work described in the Journal. Through these sources the practical requirements of the scheme are brought into sharper focus: the planning, negotiating and financing, the employment of the engineers and other workmen, and the materials and equipment that were sourced, collected or made, all of which are discussed in detail in Chapter 3. Smeaton's own drawings and plans from the Royal Society archive illustrate the Journal.

After the 1763 season the Journal ends, and the letter book tails off; the continuing story of the navigation from the end of 1763 is therefore summarised very briefly, but can be read about in more detail elsewhere[1]. The concluding section is a consideration of the importance of the Calder project over the Journal years and beyond.

1 Charles Hadfield, *The Canals of Yorkshire and North East England*, Vol 1, Newton Abbott, 1972, p 48-63.p 53.

ACKNOWLEDGEMENTS

This study stemmed from a project conducted by Wakefield Historical Society in 2012-2014 on the theme of Wakefield Waterfront. Our research for this inspired us to find out more about the origins of the navigation along which the waterfront warehouses, maltings and mills had ranged. We started to find a number of important sources, including Smeaton's journal, but our background in local history research did not extend to 18th century canal engineering, and so a quest for information and clarification began.

Our thanks go to Katherine Marshall at the Royal Society, Carol Morgan at the Institute of Civil Engineers, and the staff of West Yorkshire Archives at Wakefield, Halifax, Huddersfield and Bradford. The archivists at Claremont in Leeds, when it was the home of the Yorkshire Archaeological and Historical Society, and the Wakefield Local Studies librarians have been most helpful. We are grateful for the advice of members of the Halifax Antiquarian Society, the Railway and Canal Historical Society, and the British Bryological Society who gave advice on the use of moss.

We have benefitted greatly from the generous explanations and suggestions from members of the Newcomen Society with regard to early canal engineering methods, the equipment used, and contemporary illustrations. Mike Clarke's background in the engineering of Pennine waterways and his willingness to share his expertise and sources have been invaluable. Our thanks are due to Victoria Owens for sharing her research on Brindley's notebooks and her help with canal terminology. The willingness of Victoria Owens and Mike Clarke to read through our manuscript, and make suggestions and comments was appreciated. Christine Richardson, Sarah-Jane Stagg and Mike Chrimes also offered help and suggestions.

Our friend, Andy Beecroft, an experienced local civil engineer, who in retirement is a volunteer working on the maintenance and restoration of the present navigation, has been indispensable in guiding us towards an understanding of Smeaton's drawings, vocabulary and materials. Suggestions from Richard Bell about local sandstone, and from Alan Betteridge about the Halifax Turnpike were appreciated. Our thanks to Kevin Young who improved the quality of some of the illustrations.

David Scriven, a member of Wakefield Historical Society, was kind enough to read through our manuscript and offer suggestions; his considerable experience as a local historian was appreciated. Phil Judkins from Wakefield Historical Society encouraged us in structuring our writing.

Our thanks are due to our patient publishers, Wakefield Historical Publications, to Paul Buckley of Riasca for typesetting and to Jody Ineson for indexing.

ABBREVIATIONS

WYJS/CA	West Yorkshire Joint Services, Calderdale Archives
WYJS/WA	West Yorkshire Joint Services, Wakefield Archives
WYJS/BA	West Yorkshire Joint Services, Bradford Archives
RS	Royal Society
LI	Leeds Intelligencer
LM	Leeds Mercury
YAHS	Yorkshire Archaeological and Historical Society, Brotherton Library, Leeds University
BL	British Library
TNA	The National Archives
JHC	Journal of the House of Commons
JHL	Journal of the House of Lords
HAS/T	Halifax Antiquarian Society Transactions
WYJS/CA MIC2/16	John Smeaton's Journal 1760-1763
WYJS/CA MIC2/1	Committee Minutes, Letters and Subscriptions 1756-1758
WYJS/CA MIC2/2	Commissioners' Minutes 1758-1858
WYJS/CA MIC2/4	Copies of Reports and Letters 1758 - 1797
WYJS/CA MIC2/22	Journal, 1760 - 1765 (includes accounts)
Skempton	A.W. Skempton, (ed), *John Smeaton FRS*

CHAPTER 1

TOWARDS AN ACT OF PARLIAMENT FROM 1740

The Demand for a Navigation

> All Nations agree, that Commerce is the only means to render a State flourishing and formidable to its Neighbour ... and since by foreign Navigation it procures so great Advantages, it is natural to infer, that by INLAND NAVIGATION it must produce, in small, the same Effect ...
>
> Canals render carriages and Beasts of burthen less necessary that they may be better employed in Tillage and Agriculture: ... it is by them that Traffic can animate all parts of the State, and procure plenty and happiness to the People, and thus extend a Sovereign's Power. In fine, by Canals we can readily be supplied with Grain, Forage, Fuel, Materials for Building, and in one word, all heavy Materials, which remain of little Value 18 or 20 Miles from the Place they are wanted, because of the great Expence commonly attending their transportation by Carriages.'[2]

Before gradual road improvements began to have impact in the second part of the eighteenth century, inland waterways, coastal and other shipping routes were essential for transport and travel. Roads were roughly made, poorly drained and often impassable particularly in the steep valleys and windswept moors of the upper Calder. With the growing importance of woollen cloth-making in the area, and of Halifax as a market centre for a large region, the River Calder had early been seen as a potential water route. In 1621, 1626 and 1698 petitions were introduced to Parliament requesting that the Calder

---

2 Charles Vallancey, *A Treatise on Inland Navigation, or, the Art of Making Rivers Navigable, of Making Canals in all Sorts of Soils, and of Constructing Locks and Sluices*, 1763, Dublin. His text was extracted from the works of Michelini, Castellus, and Belidor.

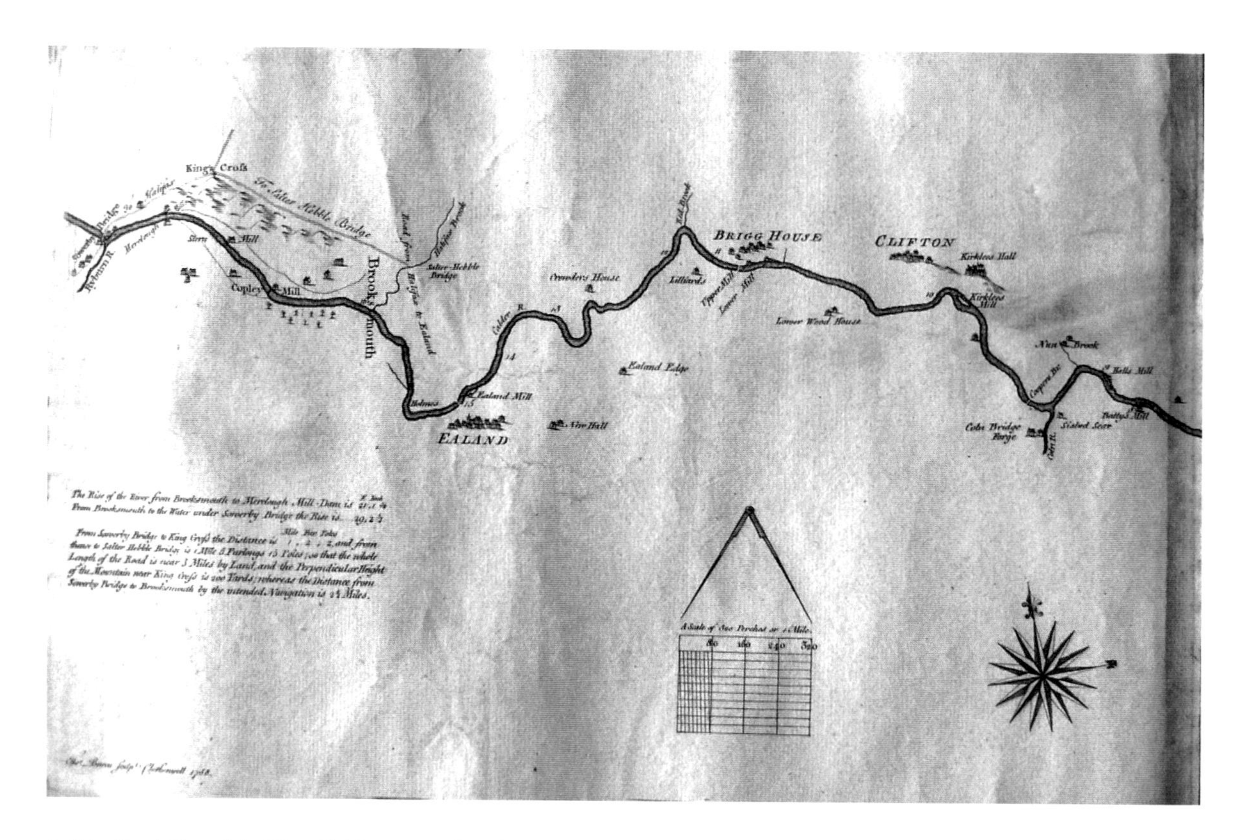

2. John Eyes' Plan of the River Calder, surveyed in the years 1740, 1741, and 1758

be made navigable to Halifax, but these proposals were blocked by the other main towns in the region. Nevertheless, at the beginning of the eighteenth century, the Aire and Calder Navigation was opened to Leeds and Wakefield.[3] In 1738 the idea resurfaced again when the merchants, clothiers, landowners and professional men of the Upper Calder Valley were considering a stretch of new navigation to Elland.[4] These promoters knew that such an undertaking would be expensive and difficult because the Calder in this Pennine area was a volatile upland river, liable to sudden and great changes in volume and flow, unlike the wider slower waterway below Wakefield leading into the Aire and the Humber.

A survey to assess the viability of this challenging scheme was commissioned from Thomas Steers of Liverpool (c1672 to 1750) and John Eyes (d 1775). The former had gained a considerable engineering reputation in projects connected with water, from the

---

3 John Hargreaves, *Halifax*, Edinburgh, 1999, p 33.

4 YAHS Collection, MD 335/3/10, No 162b, 29th Jan 1738, Draft letter from Cavendish Nevile to unspecified 'Lord'.

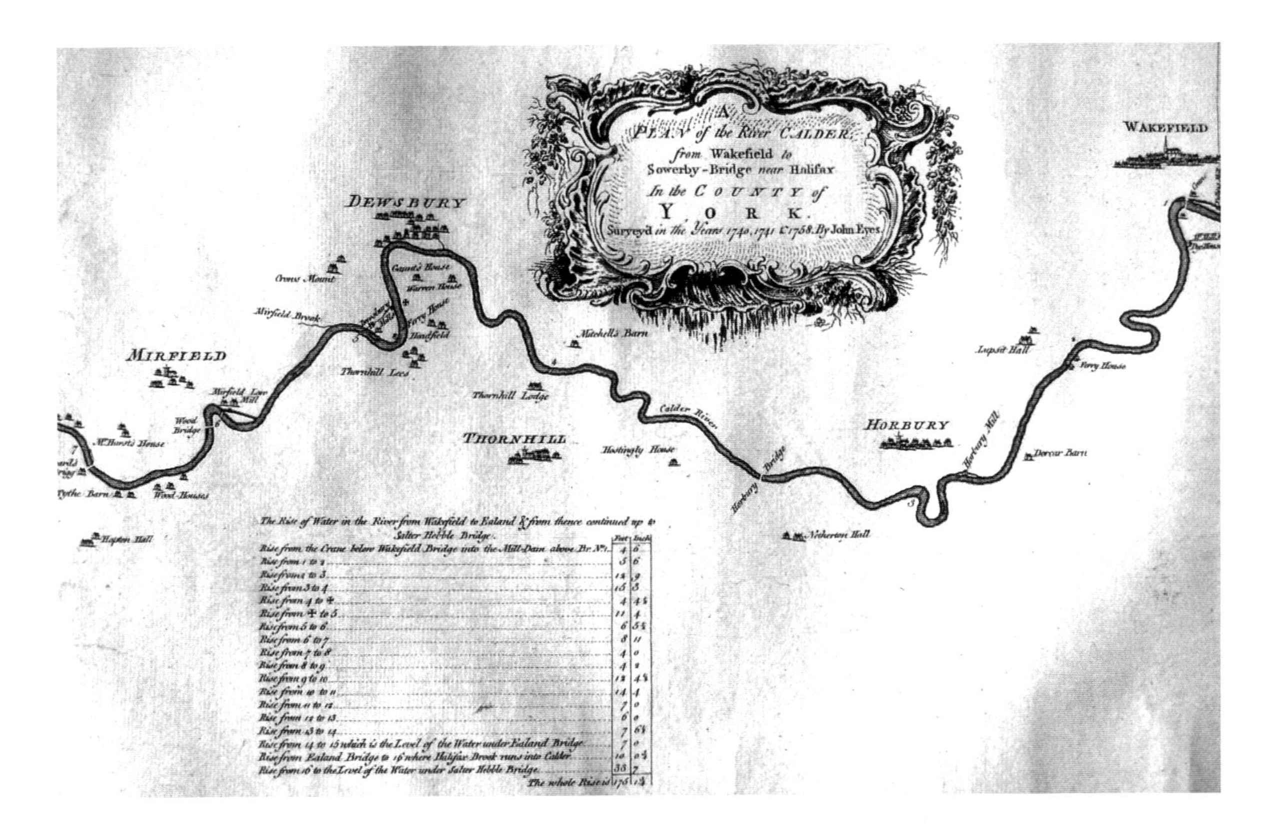

The Rise of Water in the River from Wakefield to Ealand & from thence continued up to Salter Hebble Bridge.

	Feet	Inch
Rise from the Crane below Wakefield Bridge into the Mill-Dam above Br. No. 1	4	6
Rise from 1 to 2	5	6
Rise from 2 to 3	12	9
Rise from 3 to 4	16	3
Rise from 4 to +	4	4½
Rise from + to 5	11	4
Rise from 5 to 6	6	5½
Rise from 6 to 7	8	11
Rise from 7 to 8	4	0
Rise from 8 to 9	4	2
Rise from 9 to 10	12	4½
Rise from 10 to 11	14	1
Rise from 11 to 12	7	0
Rise from 12 to 13	6	0
Rise from 13 to 14	7	6½
Rise from 14 to 15 which is the Level of the Water under Ealand Bridge	7	0
Rise from Ealand Bridge to 16 where Halifax Brook runs into Calder	10	0½
Rise from 16 to the Level of the Water under Salter Hebble Bridge	33	7
The whole Rise is	176	1½

planning of Liverpool Docks to schemes for navigations on the River Irwell, the River Douglas and the River Weaver. By 1739 he was Mayor of Liverpool. John Eyes worked with him as surveyor in Liverpool and elsewhere.[5] Thus these two men brought considerable experience in the surveying and construction of navigations west of the Pennines to their evaluation of the Calder development. Their conclusion was that the River Calder from Wakefield to Elland was 'capable of being made navigable' and their plan showed how this could be achieved. In addition a continuation by way of Brooksmouth to Salter Hebble Bridge and another up Halifax Brook into Halifax was, they felt, possible. Altogether this would bring a navigation a further twenty-four miles upriver from Wakefield.

Such a major infrastructure development needed the approval of a Parliamentary Act to provide the authority to push through the scheme by framing the regulations for managing the project, raising finances, and settling disputes. Thus a petition from Halifax was presented to Parliament on 9th December 1740 from: 'the Gentlemen, Merchants,

---

5 http://www.mikeclarke.myzen.co.uk/ThomasSteers.htm. The Douglas Navigation. Mike Clarke, Founder of the Leeds & Liverpool Canal Society.

Clothiers and other Inhabitants of the Towns, Townships, Chapelries and Precincts of Halifax, Ealand and Ripponden setting forth that the River Calder was heretofore made navigable by Act of Parliament to the Town of Wakefield, and is capable of being made further navigable to the aforesaid Towns of Ealand and Halifax, and … that this … will be highly advantageous, to all the Clothing Towns, and Places adjacent.'[6] The argument was that it would also benefit these communities by taking the pressure off the roads, which carried large and increasing volumes of traffic and were expensive to maintain. Because of the Pennine terrain the access into Halifax from all directions had long presented major difficulties with roads often unpassable for months at a time in winter. The two great travel writers of this pre-turnpike era, Celia Fiennes and Daniel Defoe, both made particular comment: when Fiennes was in the area in 1698 she never visited Halifax, the road being 'so stony and difficult'[7]; in the 1720s Defoe expressed great surprise that such a busy town could be in such an inhospitable situation with hills which made 'coming and going exceedingly troublesome, and indeed for carriages hardly practicable.'[8] Little had improved by the 1730s, the land carriage of supplies of wool and corn from the navigation heads at Leeds or Wakefield remained expensive and variable, dependent on the weather and the season. In comparison water transport appeared cheap, reliable and convenient: prices of 'merchandizes and commodities' from London and elsewhere would fall if brought up from Wakefield to Halifax by water.

Two other petitions supporting the scheme were presented later in the Parliamentary process, the first from the merchants, traders, manufacturers and inhabitants of Manchester and the second from those of Rochdale parish and thereabouts. Rochdale, in particular, was really part of the Yorkshire cloth manufacturing, receiving the coarser wool from Halifax and producing, according to Defoe, a 'sort of coarse goods called half-thicks or kersies'[9]. Like the petitions from West Yorkshire they argued that the cost of the carriage of their manufactures was high. These petitions from Lancashire indicate the considerable amount of traffic passing over the Pennines and the strong interest there in this potential waterway in the Eastern Pennines.

Witnesses at the Parliamentary Committee which was set up to consider these three petitions provided more detail about the advantages perceived. Eyes, the surveyor, had calculated the potential difference in the costs of the carriage of goods by land or on water: 'the present price of Land Carriage betwixt Halifax and Wakefield being 15sh per Ton: which might be reduced by Water Carriage to 9sh a Ton, Lock Dues included.' A second witness explained that land carriage had to be paid in both directions from the navigation

---

6 JHC, Vol 23, 1740.

7 Ed Christopher Morris, *The Illustrated Journeys of Celia Fiennes*, London, 1982, p 182

8 John Hargreaves, *Halifax*, Edinburgh, 1999, p 33.

9 Herbert Heaton, *Yorkshire Woollen and Worsted Industries*, Oxford, 1920, p 122, p 286, p 257.

and, for the clothiers, this was increased by the need to re-pack cloth between road and water at Wakefield or Leeds. He believed that the price of carriage for wool could be reduced by a third and that of wheat from Wakefield halved.

The extremely large population dependent on goods carried into this upland area for both food and work was emphasised: 'above 60,000 Inhabitants of the town and parish of Halifax are employed in the Woollen Manufactury, and in the several parishes adjacent above double that number: ... this great Manufactury would be much benefitted by continuing the Navigation ...'[10] Defoe had noted in the 1720s this huge and industrious population spread around the parish of Halifax who 'sow little and hardly enough to feed their poultry' and who 'must necessarily have their provisions from other parts of the country'.[11] The Halifax historian, John Hargreaves, has calculated population figures based on Visitation Returns which suggest that the Halifax town and parish contained 31,000 people in 1743 and over 41,000 in 1764, a smaller number than presented to the Parliamentary Committee but still very remarkable.

After hearing the Committee's report, leave was granted to bring a bill to Parliament. Miles Stapleton and Chomley Turner, the Yorkshire MPs, were to prepare it for the end of January.[12] Navigations were in fashion and there was optimism about the progress of the bill.

However, opposition to this extension of the navigation had been emerging since the scheme was proposed. At the centre of this was the owner of the Wakefield Soke mills, the Reverend Cavendish Nevile of Chevet Hall, situated between Wakefield and Barnsley. A collection of letters to Nevile or draft letters by him relate to this issue and show the sorts of discussion and lobbying from Wakefield which went on to build opposition to the bill and present the case against the extension of the navigation in Parliament.[13] The overall argument was that: 'It will be very detrimental to our landed interest in General, to our trades and markets, for ten miles round the neighbourhoods of Leeds, Wakefield and Barnsley.'[14]

Nevile was a landowner himself, addressing the idea of a threat to the 'landed interest'. The letters from the local gentry indicate the work he was doing to raise awareness amongst them of the problems he envisaged and to mobilise their support. These correspondents included Sir Lionel Pilkington of Stanley, his brother-in-law, and Lady Strafford of Wentworth Castle, recently widowed and with substantial property interests in Wakefield.

---

10 Hargreaves, *Halifax*, p 68.

11 Ed Pat Rogers, *Daniel Defoe, A tour through the whole Island of Great Britain, Exeter,* 1989, p 176.

12 JHC, Vol 23, 1740.

13 YAHS, MD 335/3/10.

14 YAHS, MD 335/3/10 No 162b Letter from Nevile to an unspecified Lord.

During the consideration of the bill other local men like Sir John Kaye of Denby Grange came out against the bill. Lord Irwin of Temple Newsam wrote to Nevile after the event to say that he had opposed the scheme 'on your account as having the greatest property that wou'd be affected'.[15] This need for solidarity between members of the gentry and aristocracy carried much weight. The 'inferior' quality of the Halifax petitioners was expressed by Nevile in a draft letter to the MP, William Wentworth of Bretton, who had been hoping to change Nevile's mind as the bill was being considered in Parliament.

'I know of no ignorant women, Carriers, or Cottages apprentices that have sign'd our Petitions. Theirs, several of the most Vulgar Sort have. Our merchants at Leeds and Wakefield and Birstall are of as good account as those of Rochdale and Manchester and those few of Halifax.'[16]

A list of the key navigation promoters had been produced for Nevile with descriptions of their standing: two 'large dealers' from Lancashire - one from Rochdale the other from Bury, four 'large dealers at Soyland and men of good substance', and 'four large dealers and men of good substance in Halifax' with three attorneys - one from Halifax, one from Ealand and the other from Almondbury. Most of these would have been very substantial men of trade and business, but not with the estates or pedigree to elevate them into the 'landed interest'.[17]

Irwin also noted that the House had been swayed by the threat to 'our trade in general'. The gentry and aristocracy, many of whom were Members of Parliament, were taking an increasing interest in the wealth of the trade of their local areas. Merchants were beginning to work closely with the older landowners as magistrates and in the Lieutenancy, and themselves becoming propertied gentlemen, investing their fortunes in land. The prosperity of towns like Wakefield, which was long-established as a centre for administration, leisure, retail, craft and professional services, all useful to the upper echelons of society, should be protected against the threat that the benefits of trade might move elsewhere. Thus the first opposition petition presented to Parliament on 11th February 1741, was mainly from 'the gentlemen freeholders farmers and others concerned in the landed interest' in Wakefield and towns and villages to the west and south of the town which would not benefit from the navigation: 'if the Bill, … as it now stands, should pass into Law, the same would be very prejudicial to the Clothing trade, and consequently to the Landed Interest in and about the several towns before mentioned.'

The owners and users of mills, of which there were a number on this section of the upper Calder, were traditionally opposed to navigations as all milling depended on the

---

15 YAHS, MD 335/3/10 No 187 H Lowther assisted Nevile by compiling 'A list of the key Undertakers of the Navigation' so that he could understand the opposition better.

16 YAHS MD 335/3/10 No 18 Letter from Nevile to William Wentworth, 24th Feb 1740.

17 YAHS MD335/3/10 No 187 Letter to Mr Nevile.

maintenance of a head of water, whilst the users of the navigation needed to release water to allow the boats to be carried through the locks. For Nevile there was an added complication in that he owned the Wakefield Soke mills. The Soke meant that the people of Wakefield parish were obliged to take their corn to his mills and he was obliged to mill it within a certain period. Nevile could be held legally liable to the Soke for his failure to fulfil his part of the bargain.[18]

To make the possible loss of water more tangible Nevile needed to demonstrate that water supply for mills could often be very badly affected by navigations. He did this by presenting evidence from businesses on the recently built Don Navigation. Certificates were prepared to send to London with this theme, for example that from John Cooke at Kilnhurst Forge near Rotherham which stated:

'a very considerable part of the Water which passes through the Navigation upon the River Dunn is lost through the Locks and Shuttles, ... and the Roguary of Boatmen leaving the Shuttles running when they go down the Cutts.'[19]

Other strong arguments about loss of water came from various opposition petitions organised by Nevile to send to London. Two petitions presented on February 12th came from 'merchants, traders, and dealers in cloth', the first from traders in the Calder Valley from Wakefield to Brighouse, and the second from those of the Leeds area sharing the same concern that the navigation would affect the fulling of cloth: 'Taking from the present Fulling Mills, on the River Calder, so much of their Water, that ... the petitioners will not only be incapable to execute the Orders for Goods ... but likewise the cloths ... will be greatly damaged by the frequent Hindrances and Stoppages that will thereby happen'. Nevile expressed himself very forcefully in a private letter on this subject: 'to extend the navigation would ruin all the Fulling mills, where their coarse and white cloth is made nine parts out of ten ... tis impossible a Cloth Trade and a Navigation upon one and the same River can Subsist.'[20]

For all sections of the West Riding woollen trade the fear that orders of cloth might not be met would have carried great weight. In particular the danger of the loss of the huge 'Russia Order', which supplied the Russian Army, was a possibility suggested by John Milnes, the influential head of Wakefield's leading merchant family, and Mr Fenton, agent for Lord Strafford. However, in the Parliamentary Session of 1740/41 Milnes preferred to use his undoubted influence to promote new turnpikes. Perhaps his reticence was to protect his relationships with many of his cloth suppliers, the Pennine clothiers.[21]

---

18 YAHS, MD 335/3/10 No 166 Letter to Nevile from Lawyer Cookson of the Middle Temple. Probably serving the Leeds interest, he supplied Nevile with advice.

19 YAHS, MD 335/3/10 No 180 From John Cooke at Kilnhurst Forge near Rotherham 22nd Feb 1740. No 182 From Mr Cotton Ironmaster at Haigh, 23rd Feb 1740.

20 YAHS, MD 335/3/10 No 186 Letter Nevile to William Wentworth 24th Feb 1740.

21 YAHS, MD 335/3/10 No 176 Letter from Alan Johnson to Nevile 19th Feb 1740.

Criticism of the old navigation, the Aire and Calder, was employed by Nevile: 'Tis very accountable why Leeds and Wakefield oppose the Bill, because they find great inconveniencies and Hardships from the present Navigation'.[22] Fellow landowners would appreciate the complaints he made about the Aire and Calder Navigation downstream of Wakefield. He railed against the freedom to use land for towpaths and to make gates, a freedom which was not recompensed sufficiently for the inconvenience caused, with charges often not paid. River banks suffered from stakes driven into them to aid 'hauling' boats, and the 'horses and extraordinary number of hands to draw the great burdened vessels up in strong water' caused great damage. If gates were left open animals would stray, or be stolen by passing 'lawless'[23] boatmen. The various roads to the water along which local heavy goods like coal or lime were taken to the river, suffered great damage.

In their demand for new turnpikes to improve road transport, both merchants and workers in the cloth trade from all parts of the region were also highly critical of the monopoly of the undertakers[24] of the Aire and Calder Navigation. As things were, they said, there was not a great difference between costs of road and water transport during summer, and in winter the navigation raised prices to take advantage of its monopoly when the roads were impassable.[25] This evidence contradicted that of the navigation witnesses, Eyes and Alderton, who had simply stated that water transport would always be cheaper than the best roads.

Co-incidentally or not, a flurry of turnpike petitions for the West Riding were passing through Parliament at precisely the same time as the upper Calder Navigation bill and these offered an alternative prospect of better transport up the Calder Valley. The first stretch of turnpike road in the area from Rochdale to Elland was begun in 1735 and bills for other major turnpike routes were proposed in 1740: from Red House near Doncaster through Wakefield to Halifax, and from Wakefield to Pontefract and on through Knottingley to Weeland, near the confluence of the River Aire and the Ouse. These changes had the potential to provide a good direct road system from all the towns of the Calder valley to the Great North Road and the Humber Estuary, circumventing the old navigation. Other turnpikes too would serve other centres: in the woollen districts there were proposals for a turnpike road between Leeds and Elland and another from Halifax by way of Bradford and Leeds to Selby to access the River Ouse, providing another route into the Humber Estuary.[26]

---

22 YAHS, MD 335/3/10 No 186 Letter Nevile to William Wentworth 24th Feb 1740.

23 YAHS, MD 335/3/10 No 162b Draft Letter from Nevile to unspecified Lord 29th Jan 1738.

24 The Leeds and Wakefield promoters who were authorised by the Act of Parliament to organise the building and maintenance of the navigation and to raise money through charges for the passage of goods.

25 JHC Vol 23, 1740.

26 William Albert, *The Turnpike Road System in England 1663-1840*, Cambridge 1972, p.48.

If these had been supported by the merchants of Wakefield and Leeds in part to offer an alternative to the proposed navigation and its threat to their control of trade, this did not arise in the expression of the arguments in Parliament. There was a general desire for turnpikes in the West Riding as the state of the roads was a perpetual concern and the requirement for the parishes along the roads to mend their stretches impinged on everyone's lives. The merchants also saw the opportunity to challenge the powerful monopoly held by the Aire and Calder in carrying goods to and from the Humber.

Their Parliamentary Committee witnesses considering the Yorkshire turnpikes, William Dickenson, a well-respected surveyor in the region, and Mr John Ramsden, a merchant from Halifax, dwelt on the present poor state of the roads and the possibilities of better roads leading to increasing traffic. The road through Wakefield to Halifax was 'in so ruinous a condition, occasioned by the many heavy carriages passing through the same, loaden with manufactures ... that in the winter season, it is dangerous for coaches, carriages and travellers on horseback to pass through them'. The people of the parishes and townships along this road had spent more on these roads than was required but could not maintain them adequately in the face of such heavy usage. John Milnes, the powerful Wakefield cloth merchant, speaking as a witness for the turnpikes, provided the same assessment of the quality of the road from Wakefield to Weeland passing through Pontefract and Knottingley. Its ruinous state, particularly in winter, was caused by the carriage of heavy goods such as coal or lime, making the transport of the region's woollen cloth more difficult. He spoke with great conviction about the effect on transport costs of improved roads:

> 'The lock dues upon the Rivers Aire and Calder being very high, Manufactures of the western area as also Wool and Corn from Lincolnshire and other places can be conveyed by Land carriage on the said roads when they are passable at an easier Expense than they are now carried by water ...'

The introduction of turnpike trusts was in the early stages and the main eighteenth century improvements in road construction were yet to come so that this confident expression of their impact may have been more in hope than knowledge. However the Petitions relating to the turnpikes intended to serve the Calder Valley towns presented to Parliament on 26th January undoubtedly resonated with the Members of Parliament who were at the same time considering the proposed navigation. As the Committee hearings for the West Riding turnpikes and the navigation bill were within a few days of each other, evidence from one bill would have had a direct bearing on the other bills which were being heard. Indeed, the petitions for the Halifax to Weeland and Redhouse Turnpike were explored in committee on 10th, 12th, 16th and 17th February and the first witness against the navigation was heard on 19th February.

The majority of the letters in Nevile's bundle are from the lawyers sent to London to lobby in Parliament during the hearing of the navigation bill, in particular the young Alan Johnson of Westgate in Wakefield, son of the Deputy Clerk to the West Riding. He evoked the drama and excitement as arguments and counter-arguments gradually emerged and perceptions altered during the days after the first petition. He reported the impact of the evidence of Mr Bartram, Nevile's mill manager, when he was asked in committee on 19th February whether he thought that the present navigation was a service to trade. His reply was: 'that he apprehended it was not, because there were Bills depending in Parliament for making Land-roads to avoid 'em', Johnson noted that this 'answer seemed to operate on the Committee like Repartee, and made them look arch ...'. If the benefit of a navigation was being thrown into doubt, causing bitter disagreement between West Riding communities, and dividing the opinions of Members of Parliament, would the turnpikes not be able to solve the problems in a different way with the approval of all sections of society?[27] Nevile could summarise these arguments more stridently in his private correspondence: 'The present Navigation is an intolerable burden to the Country, no advantage to Trade, but what might be as Cheap had, and sooner perform'd by Turnpikes ... and which would serve Halifax, Rochdale and Manchester as well as our Neighbourhood, ... and Halifax may Bail at home and send them by Wagons or Carts sooner or as cheap.'[28]

At the end of January Lady Strafford reported from London: 'I have consult'd with several Members of Parliament, particularly Sir Miles Stapyleton who seems to be of the opinion the bill will not pass, as so many Objections have been made to it.'[29] However the Wakefield lobbyists and witnesses in London continued uncertain about the outcome, especially as the presentation of their petitions was delayed until the last moment. On 19th February Johnson wrote to Nevile 'notwithstanding that Navigations are favourites there did not appear to be any one insensible that we have several Objections to this worthy of Consideration; and I can't help thinking that the Bills which are depending for the Turnpikes, are applicable to our Case so as to be made Arguments against the Navigation.' On 24th February Bartram still thought the Navigation Bill would be passed, 'the Majority are for the River being made Navigable And are begun to Examine the Bill, but they Seem to differ about several things therein as if they wou'd not get thro it hastily.'

According to Johnson it was a strategic plan to present the opposing petitions and their witnesses as late as possible in order to keep the hearings going until the end of the current Parliamentary Sessions. It seems that this policy worked as the session time to examine evidence and resolve disagreements ran out. Meanwhile the turnpike bills

27 See petitions for turnpikes eg Redhouse to Halifax, Journal of House of Commons 1740 Vol 23 p 573 'Gentlemen, clergy, tradesmen, freeholders of Wakefield and many others Justices of the Peace, Gentlemen, clergy, freeholders, tradesmen and farmers of the said Riding' who signed a petition presented to Parliament on 8th January.

28 YAHS, MD 335/3/10 No186 Nevile to Wentworth 24th Feb 1740.

29 YAHS, MD 335/3/10 No 163 Lady Strafford to Nevile 28th Jan 1740.

became law by 21st March 1740. For the merchants and clothiers of the upper Calder this was a huge disappointment which set their campaign back a number of years. In the late 1750s when the scheme was revisited the Halifax promoters interpreted their 1740 failure:

> 'Some Gentlemen opposed it, and insisted that they had not had reasonable Notice of the then intended Project, nor any Opportunity of examining the Plan proposed, so as to be able to judge how far their Estates adjoining to the said River, might or might not be affected by it. It was therefore, at that Time thought proper to postpone the further prosecution of the said Navigation till such Time as every Person interested might have Opportunity to consider thereof, and how far the same might or might not be serviceable to the Public.'[30]

It seems that in 1740 the navigation promoters from the upper Calder had been taken by surprise by Nevile's vigorous opposition campaign, and their arguments had suffered because the West Riding turnpike evidence had also been heard over the same period. They needed to build their confidence and find wider and more powerful support to launch a new campaign, dealing with all possible areas of disagreement to ensure that next time their considerable investment of time and money would be successful.

Working towards a new Navigation Act

There appears to have been another failed attempt to produce a new Parliamentary petition for the extended navigation in 1752 when several Halifax meetings were held and a Mr W Turner of Blake Hall, Mirfield, was requested by John Caygill of Halifax and Robert Allison of Ripponden to note 'all the Reasons that I could imagine to be for or against the Matter then in Question'. He had sent the 'Reasons and Answers' he had devised to John Stanhope of Horsforth who with Mr Wilson, both lawyers, would 'put them into form' for publication. Although Stanhope thought Turner's work satisfactory, the new impetus seems to have petered out, and Turner was never paid for his trouble.[31]

However, on April 2nd 1757 a group of interested gentlemen met at the Talbot Inn in Halifax to 'consider proper measures to obtain an Act of Parliament for making the River Calder navigable from Wakefield to Elland' and on to Halifax.[32] This venture immediately looked very promising: the leading name on a list made that day to establish a committee was that of Sir George Savile, (1726-1784), 8th Baronet of Thornhill, grandson of the Reverend John Savile, rector of Thornhill. Although he inherited Rufford Abbey in Nottinghamshire, and had a London town house, he spent considerable time in Halifax, and at his death

30 WYJS/BA SpSt/13/2/2, *Reasons for extending the Navigation of the River Calder from Wakefield to Halifax*, August 1757

31 LI 19th Oct 1756. Mr Turner was appealing for payment.

32 WYJS/CA MIC2/1 2nd April 1757. LI 30th Aug 1757.

was buried in the family vault at Thornhill.[33] He was a Whig, with liberal views and gained the support of both Tories and Dissenters when he was elected unopposed as one of the two Members of Parliament for Yorkshire in 1759: he remained an MP until 1783. He had been a member of the Royal Society from 1747. Under his auspices the Union Club was established in Halifax, meeting at the Talbot. The Club was an association of the gentry and tradesmen of the area formed to express their loyalty to the House of Hanover. It was as a result of their meetings that the new proposals appeared and their members became the first committee members. Savile owned a considerable acreage on the north side of the Calder through which a new navigation would pass. Highly respected locally and with considerable influence and connections he would have been indispensable in driving forward the project both in Parliament and in the region.

3. The Talbot Inn from 'Sketches of Old Halifax' by Arthur Comfort, Halifax Courier, 1911-1912

4. Sir George Savile holding the Calder plans, by Basire after Benjamin Wilson, published 1770. © National Galleries of Scotland

The second reason to be optimistic was the standing of the Halifax gentlemen who attended the meeting. As leading citizens of the town, their prosperity had grown as Halifax developed to become the most important worsted manufacturing centre in the West Riding and a major centre of commerce by the mid-century.[34]

33 http://www.historyofparliamentonline.org/volume/1754-1790/member/savile-sir-george-1726-84.
34 John Hargreaves, Halifax, p. 29.

They would already have known each other well from the Union Club, through their business interests and a number of family intermarriages. John Caygill (c1708-1787) who was later appointed joint Treasurer by the navigation commissioners, owned The Shay, Halifax, which was inherited through his mother, Martha née Stead. She was related to Valentine Stead whose son, another Valentine, was also a regular attender at meetings of the committee. John's father had bought land at the Square in Halifax in the 1750s and John built brick houses on it designed by John Carr. In 1779, he and his wife gave land at Talbot Close and a sum of 800 guineas for the construction of the Piece Hall, still Halifax's most imposing building.[35]

5. Portrait of John Caygill by John Hoppner

David Stansfeld (1719/20-1769) was described as a merchant when he gave evidence on the proposed Calder Navigation to the House of Commons Committee.[36] During the years covered by the Journal, he purchased the Hunger Hill Estate towards the south of Halifax from Jeremiah Rawson and built Hope Hall there, today still a fine urban mansion though much neglected. He inherited other property, including two fulling mills at Longbotham, near Sowerby Bridge on the Calder. With Dr Cyril Jackson MD (1717-1797) he acted for the committee to deal with correspondence, arrange meetings and prepare newspaper advertisements. Dr Jackson, too, lived in a large house to the south of Halifax, Upper Calico Hall; he married another member of the Rawson family, Judith, in 1744.[37]

The Royds family were later to act as treasurers for the commissioners with Caygill. Jeremiah Royds (c1710-1786) of London, was in partnership with his brothers: John, a wool and cloth merchant and banker, and Robert.[38] In 1766 John, (c1720-1781),

35 HAS/T 1961, '*The Square and Piece Hall*', R. Bretton. Malcolm Bull's *Calderdale Companion*, http://www.calderdalecompanion.co.uk

36 31 Geo II JHC Vol 28.

37 Malcolm Bull's *Calderdale Companion*. http://www.calderdalecompanion.co.uk.

38 Sir Clement Molyneux Royds, *The Pedigree of the Family of Royds*, London, 1910.

6. Hope Hall, Halifax, built by David Stansfeld about 1765

built a 17-bay mansion in George Street, Halifax, designed by John Carr, part of which remains and is now known as Somerset House.[39] Other dedicated supporters of the project included William Gream of Heath Hall, Halifax, who was a Captain in Sir George Savile's Battalion of Militia,[40] and William Prescott (1720-1791) of Calico Hall, Halifax (later known as Clare Hall). William's younger brother, John Prescott (1726-1795) also attended regularly; he was a merchant and woolstapler in Halifax and was in partnership with the Royds family. Their sister, Ann, married the town's leading attorney, Robert Parker, who was in partnership at this time with another regular committee member, the attorney William Baldwin.[41]

These were the names that are to be found as regular attendees at meetings of the committee and these men later acted as commissioners and subscribers to the new navigation. Other substantial Halifax people attended the commissioners' meetings sporadically, and were joined by a small group of men from Rochdale. With the guidance of Sir George Savile, they were the driving force of the project.

The final reason for confidence was that the first decision of the Halifax gentlemen on April 2nd was to ask John Smeaton to become involved in their plans, initially to come to view

7. Royds House, Halifax. The remaining portion is now known as Somerset House

---

39 HAS/T 1941, *Royds of George Street Halifax and of Bucklesbury London*', T.W. Hanson.

40 *The Monumental and other Inscriptions in Halifax Parish Church*, Ed. E.W. Crossley, Leeds:1909.

41 HAS/T 1967 '*Clare Hall Halifax*', C.D. Webster.

the river.[42] Smeaton (1724-1792) grew up at his family home at Austhorpe, east of Leeds, where, by the age of sixteen, he had his own workshop to practise his early interest in mechanics. He was well-educated but did not pursue the profession of a lawyer which his father favoured, instead opting out of legal training after two years in London from 1742-44 to return to Austhorpe and continue his experiments in mechanics, designing apparatus for electrical experiments and developing an interest in astronomy. To further his study of engineering and science he was also learning French. In 1748 he was once more in London where he set up a successful business as an instrument maker. By 1751 he was able to expand into larger premises and he began to conduct experiments on the power of wind and water using working models. His skills were recognised by the members of the Royal Society and, in 1753, aged only twenty-eight, he was made a Fellow:

8. John Smeaton about 1759 by Benjamin Wilson

> 'for his great skill in the theory and practice of Mechaniks (of which he has given repeated proofs in the papers he has already communicated and the instruments he has invented) as also in being well-versed in abilities in mathematics and natural philosophy'.[43]

He began to take engineering commissions, which allowed him to understand more about the power of water and wind through practical applications. In 1753 he

---

42 WYJS/CA MIC2/1 2nd April 1757.

43 RS, EC/1752/34, John Smeaton.

was commissioned to design a watermill at Halton in Lancashire, and his second watermill was built in 1754 on Wakefield Bridge for Sir Lionel Pilkington, the new owner of the Wakefield Soke mills. Smeaton continued to research the science and application of water and wind, studying Dutch books on mills, and Belidor's 'Architecture Hydraulique'. In 1755 he spent five weeks on the Continent, landing at Dunkirk and travelling, mostly by canal, through Belgium and Holland. He inspected systems for draining the low-lying land and viewed the methods of construction used for harbours, docks and mills. On his return he visited Scotland to survey the River Clyde below Glasgow, and was consulted about fen drainage at Adlingfleet near Goole. In 1755 he was back in Wakefield designing a windmill for Mr Roodhouse of Westgate for the production of rape-seed oil and logwood for use in dyeing.[44] All his experiments, reading, and discussions, his visits and practical experience in designing and constructing mills contributed to his paper for the Royal Society, 'An Experimental Enquiry concerning the Natural Powers of Water and Wind to turn Mills.' It was presented during the summer of 1759 and the Royal Society recognised the significance of this work by awarding him the highly prestigious Copley Medal for his original research.[45]

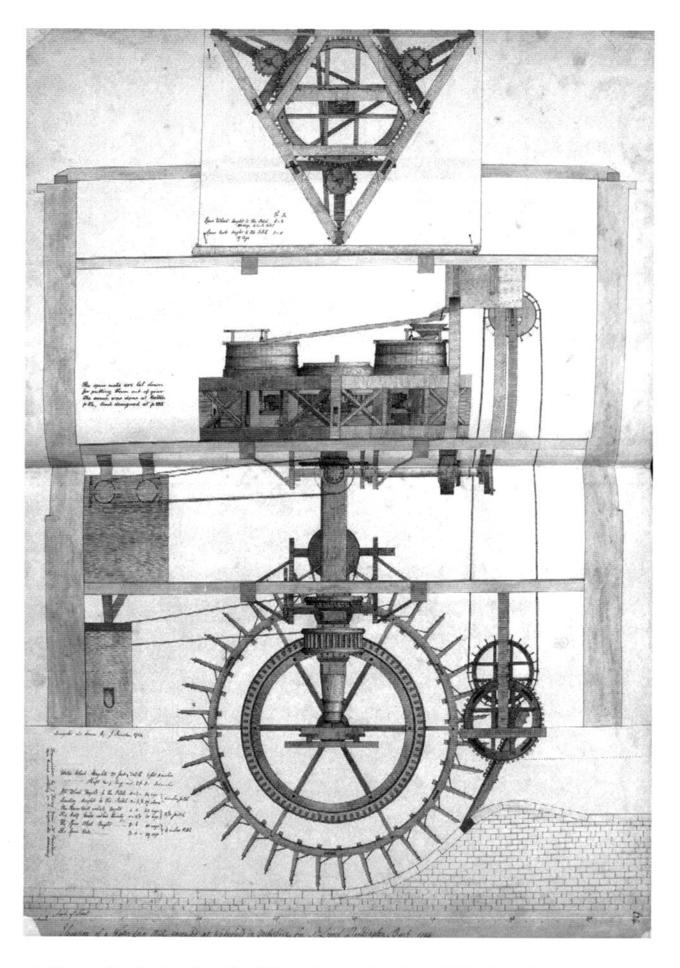

9. Part of a design for the Water Corn-Mill on Wakefield Bridge

In 1757 Smeaton prepared two plans for the water supply at the very high-profile new project, the setting up of the Ackworth Foundling Hospital. The Trustees who had invited him to submit his plans included some of the leading Whig families in the whole region in particular Sir Roland Winn of Nostell Priory, the Marquis of Rockingham of Wentworth Woodhouse, and Viscount Irwin of Temple Newsam.[46]

---

44 John Goodchild, *Aspects of Medieval Wakefield and Its Legacy*, Wakefield, 1991, p 87.

45 Skempton, Ch 1, 'John Smeaton' Trevor Turner and A W Skempton, London, 1981, pp 7-16.

46 Henry Thompson, History of Ackworth School, Centenary Committee, 1879, pp 3-4, *YAS Journal Volume 61*, 1989.

10. Design for a windmill at Wakefield for rape-seed oil and logwood for dyeing

In the second half of the 1750s he worked on the engineering project that was to earn him a national reputation. In 1756, following a recommendation from Lord Macclesfield, President of the Royal Society, Smeaton was commissioned to design a lighthouse on the Eddystone Rocks to replace the one designed by John Rudyerd, which had burned down in December of the previous year. Smeaton took the decision that the new lighthouse should be built in stone to withstand high seas, and that the building should be round to offer the least resistance. His design dovetailed the stonework into the rock for strength and stability. Construction was limited by tides and weather so careful planning was essential. Smeaton chose quality materials and paid skilled craftsmen generously. He realised that the durability of the mortar was vital and with the help of William Cookworthy, pioneer of English porcelain, he experimented with lime and pozzolana[47] to find a mortar that would set as hard as stone in wet conditions.[48] This hydraulic mortar he later used in the construction of locks and weirs on the Calder. Smeaton's work on the Eddystone gave him considerable experience in designing a structure to withstand ferocious storms. He was able to apply this knowledge later when he needed to consider methods of combatting the volatility and power of the upper Calder.

Sir George Savile is very likely to have known Smeaton personally through his work at the Royal Society: he would certainly have been aware of his research and reputation. Smeaton's particular experience of working in Wakefield on the Calder mill and the Roodhouse windmill may also have brought him into touch with some of the Halifax group as well as many of the Wakefield elite. Here was an engineer with recent experience on the Calder, whose social and academic background meant that he was comfortable

---

47 Pozzolana for the mortar was imported from Italy; this is a fine volcanic ash added to lime mortar to strengthen its durability.

48 http://www.engineering-timelines.com/who/Smeaton_J/smeatonJohn7.asp

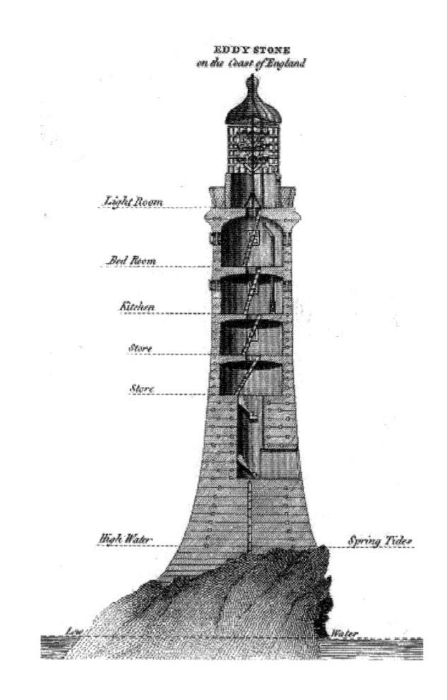

11. Contemporary drawing of Smeaton's Eddystone Lighthouse

mixing at the highest levels of society, and whose understanding of water management made him the foremost authority on the subject in the country.

Smeaton agreed to come to look at the river after the first approach in April, but for several months the committee, through their chosen representatives, Cyril Jackson and David Stansfeld, had to be content to correspond with him by letter while Smeaton was working on the Eddystone. On 23rd June 1757, Jackson and Stansfeld wrote asking him to come and survey the river as soon as he could and, in the meantime, to advise them on 'any steps that can be taken that will be of service in expediting this affair'. He responded on the 15th July that he would not be able to come until the end of August at the earliest but asked for 'a plan to be made of the river laid down to a Scale, describing its various breadths, turnings, windings which are to be done by any tolerable Surveyor of Lands, and also to procure a Light Boat ... in order to examine the Depths of the Water, Shoal, Sands etc'.[49] Eyes' map from the survey done with Steers was sent, and the committee agreed to procure a boat for Smeaton's survey.

In the letter to Smeaton that accompanied Eyes' survey the anxieties arising from past failure were clear. 'Much will remain to be done, after you have made your report and previous to an application to Parliament, into ways of answering objections, reconciling the several persons interests in the river'. They had, they told Smeaton, left the idea for 16 or 17 years so that they might 'remove every Objection which has hitherto, or can possibly with any Appearance of reason be urged against the scheme ... and to take the benefits of the latest Improvements in such matters.'[50]

By the end of August the gentlemen were impatient to launch their scheme. Advertisements were placed by the committee in the newspapers in London, York, Leeds and Manchester for several weeks declaring the intention to extend the navigation, 'provided no reasonable Cause to the Contrary shall hereafter appear', and announcing the invitation given to John Smeaton to survey it. Anyone who had objections was 'earnestly desired'

---

49 WYJS/CA MIC2/1 23rd June 1757.
50 WYJS/CA MIC2/1 22nd July 1757.

to contact the committee.[51] They also announced that they would hold a public meeting to consider Smeaton's report. Secure in the services of a renowned - if tardy - engineer, and guided by a leading Whig gentleman the Talbot committee were confident that they could answer all objections in a transparent process before submitting the scheme to Parliament.

NAVIGATION.

WHEREAS great Numbers of Gentlemen, Merchants, and Others, well ſatisfied of the great Advantages, which wou'd accrue to the Public from extending the Navigation of the River CALDER from *Wakefield* up to *Eland* and *Halifax*, propoſe to revive an Application to Parliament for that Purpoſe, (provided no reaſonable Cauſe to the contrary ſhall hereafter appear) and have appointed a Committee to meet at the Houſe of JOHN MELLIN, the Sign of the *Talbot* in *Halifax*, every ſecond Wedneſday in the Month, or oftener, if it be judged neceſſary, to conſider what Steps may be proper to be taken previous to ſuch Application. The Committee for this Purpoſe have therefore directed Mr. SMEATON to re-ſurvey the River, and make his Report as ſoon as poſſible, which however he tells them his other Connections will not permit him to do before the Month of October at ſooneſt.——In the mean Time they take this Method of acquainting the Public with their Intentions, and of expreſſing their earneſt Deſire that all Perſons who have Objections, or may not be convinced of its Utility, wou'd be pleas'd to favour the Committee with a Line, and openly propoſe their Difficulties, in order to the having them fairly canvaſſed.

So ſoon as Mr. SMEATON ſhall have finiſhed his Survey, and made his Report, proper Notice will be immediately given of a Meeting, where ſuch Report will be made public, and where it is hoped all Perſons any how intereſted will attend.

☞ N. B. Letters directed to Dr. JACKSON, or Mr. STANSFELD, at *Halifax*, will come in Courſe, and be immediately laid before the Committee.

Halifax, Auguſt 11th, 1757.

12. Meeting about the proposed navigation in the Leeds Intelligencer 13th September 1757

While the committee members waited for Smeaton to arrive in Yorkshire there were useful exchanges by letter assisting Smeaton to an understanding of the situation. At first he imagined that expense had been a principal objection in the past, but the committee replied that the previous issues which were likely to be raised again were the injury to the fulling and other mills and the waste of water from locks, as well as the greater likelihood of flooding in adjacent land. The Leeds and Wakefield merchants and dealers would object, but the strongest opposition they felt would come from the mill owners, particularly Sir Lionel Pilkington. Smeaton commented immediately that 'nothing can have happened more lucky than what I have done for him as it will remain an incontestable proof of what can be done in the Application of Water to the Uses of Life, more than was known to our forefathers'. The committee arranged for him to look at papers from their 1740/1 lawyers whilst he remained in London. Smeaton recognised there would be legitimate concerns 'things cannot be done without hurting some particular pieces of land and some particular mills to some degree'.[52]

Eventually, in late October 1757, Smeaton arrived in the West Riding and began to conduct his survey of the river. He travelled by boat to view the river itself and then by land to learn about the mills and the landowners along its length. In his later Parliamentary evidence he described this period: he had been seven weeks employed in surveying the Calder, during which he was three times up the river and spent about a month taking the measures. He was asked by the Halifax committee to have someone with him to 'minute

51 WYJS/CA MIC2/1 1st Sept 1757.

52 WYJS/CA MIC/1 11th Sept 1757.

down every Objection you hear of, and from whom, take an exact account of Landowners etc, with the Extent of their respective Properties on each Side of the River, and have in Readiness also a Certificate proper to be signed by all such of em as can be reconcil'd to the scheme.'[53] Smeaton then began to produce his report and his plan of the proposed navigation. Meanwhile the committee agreed to encourage the millowners upstream from Wakefield to look favourably on the venture by suggesting that any building materials required by the mills would go toll free. They also began to consider the tolls on heavy cargoes such as lime, stone, dung and manure.[54]

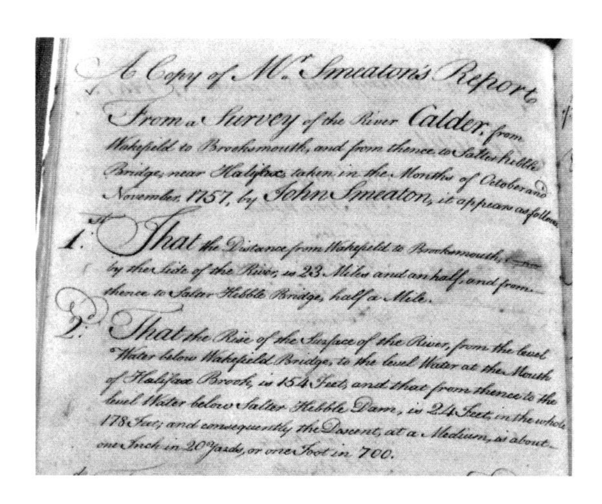
A Copy of Mr. Smeaton's Report
From a Survey of the River Calder, from Wakefield to Brooksmouth, and from thence to Salter hebble Bridge, near Halifax, taken in the Months of October and November 1757, by John Smeaton, it appears as follows

1. That the Distance from Wakefield to Brooksmouth, by the Side of the River, is 23 Miles and an half, and from thence to Salter Hebble Bridge, half a Mile.

2. That the Rise of the Surface of the River, from the level Water below Wakefield Bridge, to the level Water at the Mouth of Halifax Brook, is 154 Feet, and that from thence to the level Water below Salter Hebble Dam, is 24 Feet, in the whole 178 Feet; and consequently the Descent, at a Medium, is about one Inch in 20 Yards, or one Foot in 700.

13. Copy of John Smeaton's Report

Smeaton presented his report to the meeting of November 23rd attended by Sir George Savile. The confidence and relief of the members was palpable as they immediately began to look to the future. An abstract of the report was agreed for circulation. This was very straightforward, simply recording Smeatons's measurements and observations. The distance to be covered was 23½ miles with a rise of 154 feet. The extra section up to Salter Hebble, adding another 24 feet in half a mile, would require a reservoir. The overall fall of the river was an inch in 20 yards. He noted the mills along the route and observed that, even in the low water conditions he had experienced, they were able to work, but still with sufficient water for boats to pass. The shallow gravel or shingle beds in the river could be removed, and the meadows beside the river were of a height and gradient to avoid standing water. The conclusion therefore was that a navigation to Brooksmouth was practicable without 'sensibly affecting the Mills' or causing flooding and could take boats and barges similar to those coming up the old navigation to Wakefield, drawing 3'6". Smeaton estimated the cost as not exceeding £30,000.

The promised public meeting was to be in early December at the White Hart, one of the two main inns in Wakefield. Smeaton's report would be made public and 'all landowners or others who are any Ways concern'd are desir'd to attend'. A four-page pamphlet prepared in August entitled 'Reasons for extending the Navigation of the River Calder' was circulated

53 WYJS/CA MIC2/1 19th Sept 1757.
54 WYJS/CA MIC/1 9th Nov1757.

beforehand containing a closely argued list of nine replies to possible objections to their scheme in an attempt to forestall opposition. A thousand copies of Smeaton's report were printed. Sir George Savile advised on which gentlemen of the area should be visited by Smeaton and a member of the committee beforehand. These included Sir John Kaye of Denby Grange (1725-1789) High Sheriff of Yorkshire for 1761-2; the immensely wealthy Lord Rockingham of Wentworth Woodhouse (1730-1782) a Whig MP from 1751 and later to be Prime Minister for two short periods in the 1760s and 1780s; Lord Irwin of Temple Newsam (1691-1761) Lord Lieutenant of the East Riding of Yorkshire; Sir John Ramsden of Byram and Huddersfield, (1699-1769); and Sir George Dalston of Dalston Hall and Heath (1718-1765) MP. These were some of the most powerful and influential landowners in the region. Letters were sent to announce these deputations when 'they hope they will confirm your good Opinion of Inland Waterways in general, relieve any Fears you may entertain with respect to the general effect of this upon the Mills and would share the view that the navigation would be in the public good'.[55]

Smeaton was perhaps uniquely qualified amongst his engineering contemporaries for such visiting. He was able to use the polite discourse of the period, the civility which was designed to avoid conflict even where there was disagreement.

	The following Shoals *or* Streams *are to be dredged; and some of them weared or walled where necessary.*
a C	*From the Brook to the Tail of Horbury Cutt.*
b F	*A Stream below the Figure of 3-Lock Cutt.*
c	*A Stream below Dewsbury new Mill.*
d	*Water-gate Stream.*
e e e	*3 covered Shoals between Mirfield low Mill and Ledger Mill.*
f f	*Lyon Royd Stream.*
g g	*From below the Island to the Tail of Kirklees Cutt.*
+	*A covered Shoal in Kirklees-mill Dam.*
h h	*Willow and Lee Streams.*
i	*A covered Shoal between Brighouse and Brooksfoot Wood.*
k	*Cromwell bottom Streams.*
l	*Upper part of Long Stream.*
m X	*From the Tail of Ealand-mill Goit to the Tail of the Cutt.*
n o p	*Pighill lower Stream, Midstream, and Upper Stream.*

N.B. If the Cutts C C *and* P *are carried according to the Dotted Lines* a C *and* P g, *the dredging of the Shoals* a C *and* g g *will be avoided.*

14. Panel listing shoals and streams from John Smeaton's Plan

REASONS

For EXTENDING the

NAVIGATION of the River CALDER,

FROM

WAKEFIELD to HALIFAX.

THAT Inland Navigations conduce much to the Improvement of Trade and Manufactures, is a Poſition ſo generally acknowledged to be true, that it requires no Proof; and that the Advantages accruing from ſuch Navigations are ſtill greater in Proportion to the Difficulties or Expenſiveneſs attending Land-Carriage, in the Country where ſuch Inland Navigations are made, is a Poſition as true and ſelf-evident as the former. Now, the Town and Pariſh of *Halifax*, and the Country on each Side the River *Calder*, from that Place to *Wakefield*, is one of the moſt populous Parts of *Yorkſhire*, and full of induſtrious Inhabitants, whoſe Bread is earned in, and depends upon, the Woollen Manufacture; yet this Part of the Country, tho' full of Inhabitants, produces but a ſmall Part of the Materials uſed in the ſaid Manufacture, and ſcarce one third Part of the Corn, and other Neceſſaries of Life, conſumed by the Inhabitants therein; ſo that almoſt all the Materials uſed in the Manufacture, and above two third Parts of the Neceſſaries of Life, always have been, and ſtill are, brought from diſtant Places by Land-Carriage: But this Part of the Country is generally Moun-

15. Reasons for Extending the Navigation of the River Calder from Wakefield to Halifax

55 WYJS/ CA MIC2/4 29th Nov 1757.

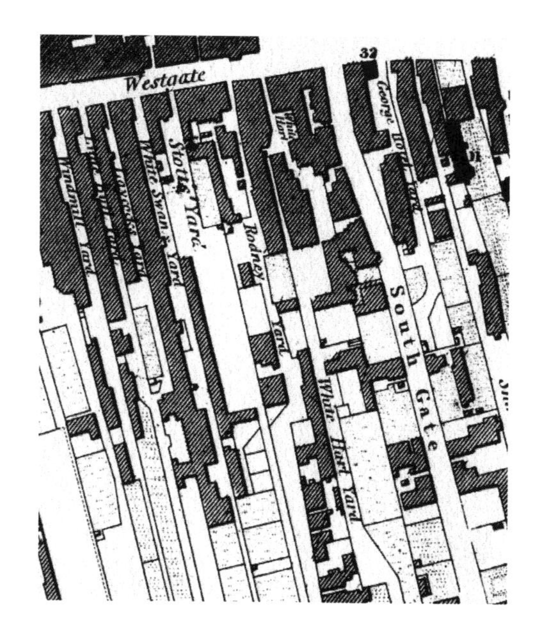

16. The White Hart in Wakefield, on the corner of Westgate and Southgate, from a map of 1823, surveyed by J. Walker

Yet Smeaton was always practical and able to communicate his findings and conclusions with great clarity. Such attributes would have been much appreciated by many of the gentlemen he visited who often interested themselves in scientific advances and improvement. The committee was ensuring that this 'landed interest' who represented the region as Members of Parliament, would consider their scheme in the best light possible.

On the 7th December 1757, at the general meeting at the White Hart, Smeaton's report was read and discussed. Present at the meeting were many of the leading gentry from the Wakefield and Halifax area: Sir Rowland Winn of Nostell, Sir George Dalston (MP for Westmoreland), Sir John Armytage of Kirklees (MP for York) as well as Sir George Savile. Two owners of Dewsbury mills, Mr Banks[56] and Mr Greenwood did raise issues but the minutes suggest these were dealt with to everyone's satisfaction. Notably, Sir Lionel Pilkington did not attend, although he had been contacted by a letter written on 29th November from the committee 'with such Answer to the Objections rais'd in 1740, as their hope may be thought satisfactory, and sufficient to relieve your Fears by Convincing you that its Effects upon the Mills cannot be so injurious as you seem'd to apprehend'. Wakefield's merchants were not present either.

On the same day a committee meeting was held in Wakefield, at which they agreed that 'no material objection had been presented.' Letters were sent to Lord Irwin, Sir Conyers Darcy (MP for Yorkshire) and Lord Rockingham to inform them that the decision had been made to petition for a Bill in Parliament. Lord Downe (MP for Yorkshire) was also informed of the decision and was asked to present the petition to Parliament and give his support.

---

56 Mr Banks is recorded in 1758 as building 'very valuable mills' near Dewsbury (Ref. John Goodchild, *Report on the History of Sands Mill Dewsbury*, YAHS/MS953 1969) He caused further problems for the Commissioners, noted in the Journal of 1763. In 1769 he claimed and won damages from the Navigation (Ref. HAS/T Chas. Clegg, Our local Canals, 1922.) Mr Greenwood owned Dewsbury Mill.

The map was not finalised until a little later. On October 22nd 1757 the committee had decided that Eyes' plan 'The Old Plate from which the Chart of the River was taken' could 'with proper alterations be made to serve the immediate purposes of the present survey'. Meanwhile Smeaton was empowered to get his new survey engrav'd upon a copper plate and to make 900 copies. On December 15th he was in London and arranged for his large plan to be engraved. He selected a Mr Seale[57] 'who was well recommended to me and was concerned in one of the neatest engravings of London and Parts adjacent that has yet appeared' and he paid £6 or £7. The beautiful engraving which resulted is a persuasive image, full of confidence. The additions of representations of the towns along the Calder and neat tables of information made his plan both attractive and informative. By January 27th 1758 copies of Smeaton's plan had been paid for. Other preparations were continuing: Smeaton was sent a list of gentlemen to canvass whilst he was in London; legal opinions about rights in the mills and lands were sought; scare stories and rumours needed to be scotched; petitions were circulated to drum up support, and supplemental subscriptions were collected.

Despite all the efforts to prevent problems, matters did not proceed smoothly. A committee of Lancashire gentlemen who met in Rochdale began to press for the navigation to be extended to Sowerby Bridge. Meetings were held to deter them from their purpose: the Yorkshire committee members were concerned that this 'must contribute no little to improve the jealousy subsisting at Leeds and Wakefield and with other gentlemen of property that the trade of the West Riding is likely to be moved Westward.'[58] By January, however, the Halifax committee needed to send their petition to Parliament to ensure the time for the passage of their bill.

The Bill in Parliament

The 'Gentlemen, Clergy, Merchants and Traders and other inhabitants' of Halifax presented their petition on January 25th 1758 stating that 'the River Calder is capable of being made navigable from Wakefield to near Halifax' and that 'extending the said river would tend to the preservation of the Highways, and the Increase of Watermen and Mariners, and be of Public Utility to this Nation'.[59] Their petition was immediately referred to a Commons committee which included Lord Downe, and Sir John Armitage, the two Members of Parliament for Yorkshire in 1758. In early February a supporting petition was presented from the small towns and villages up-river from Wakefield: 'the principal owners and occupiers of mills, lands, Tenements and Hereditaments in or near Horbury, Ossett, Thornhill, Dewsbury, Mirfield, Clifton and Hightown', and the traders

57 Richard William Seale (1732-1785) draughtsman and engraver of London.

58 WYJS/ CA MIC2/1 14th Jan 1758.

59 JHC Vol 28: 31 Geo II.

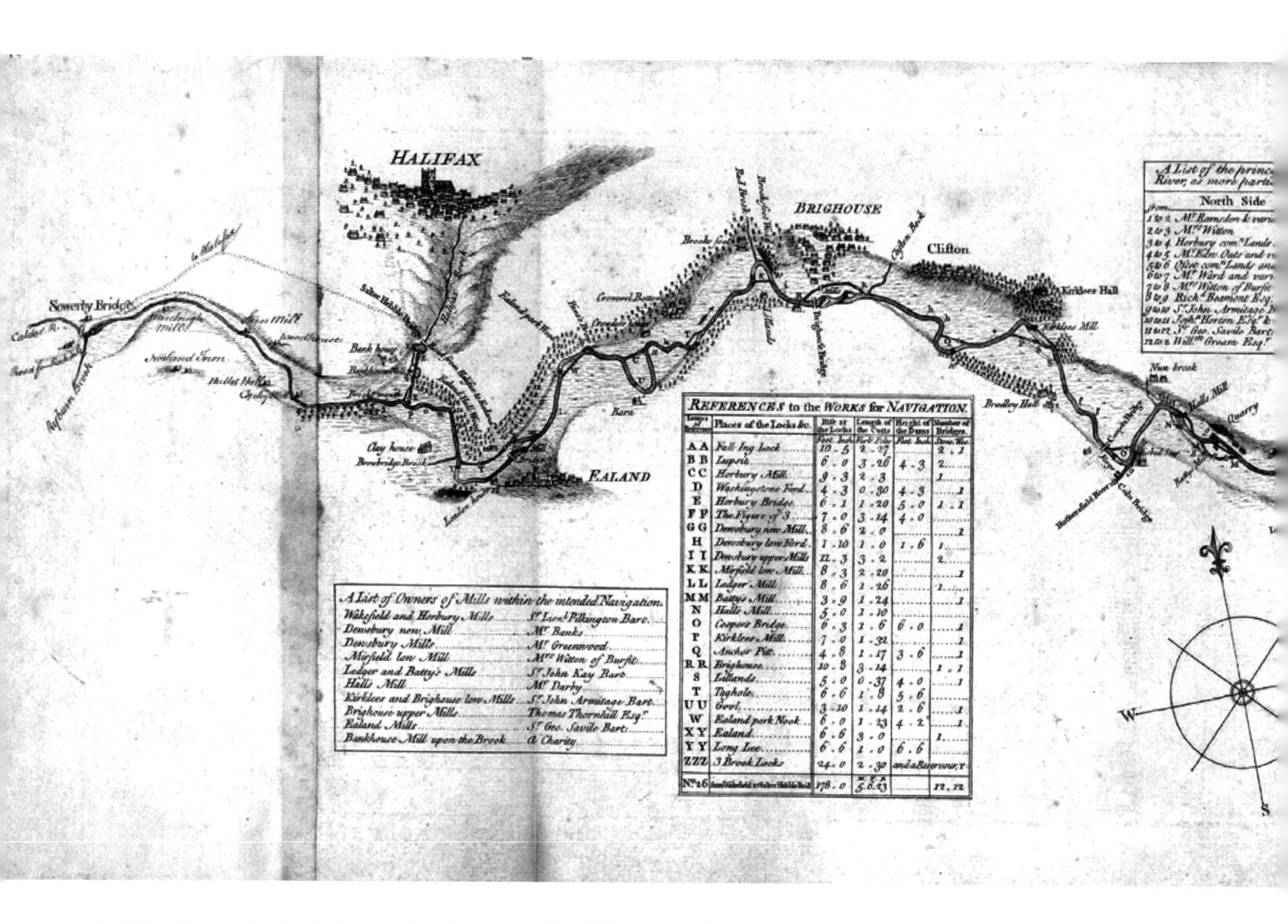

17. *John Smeaton's Plan of the River Calder, surveyed in 1757, engraved in 1758*

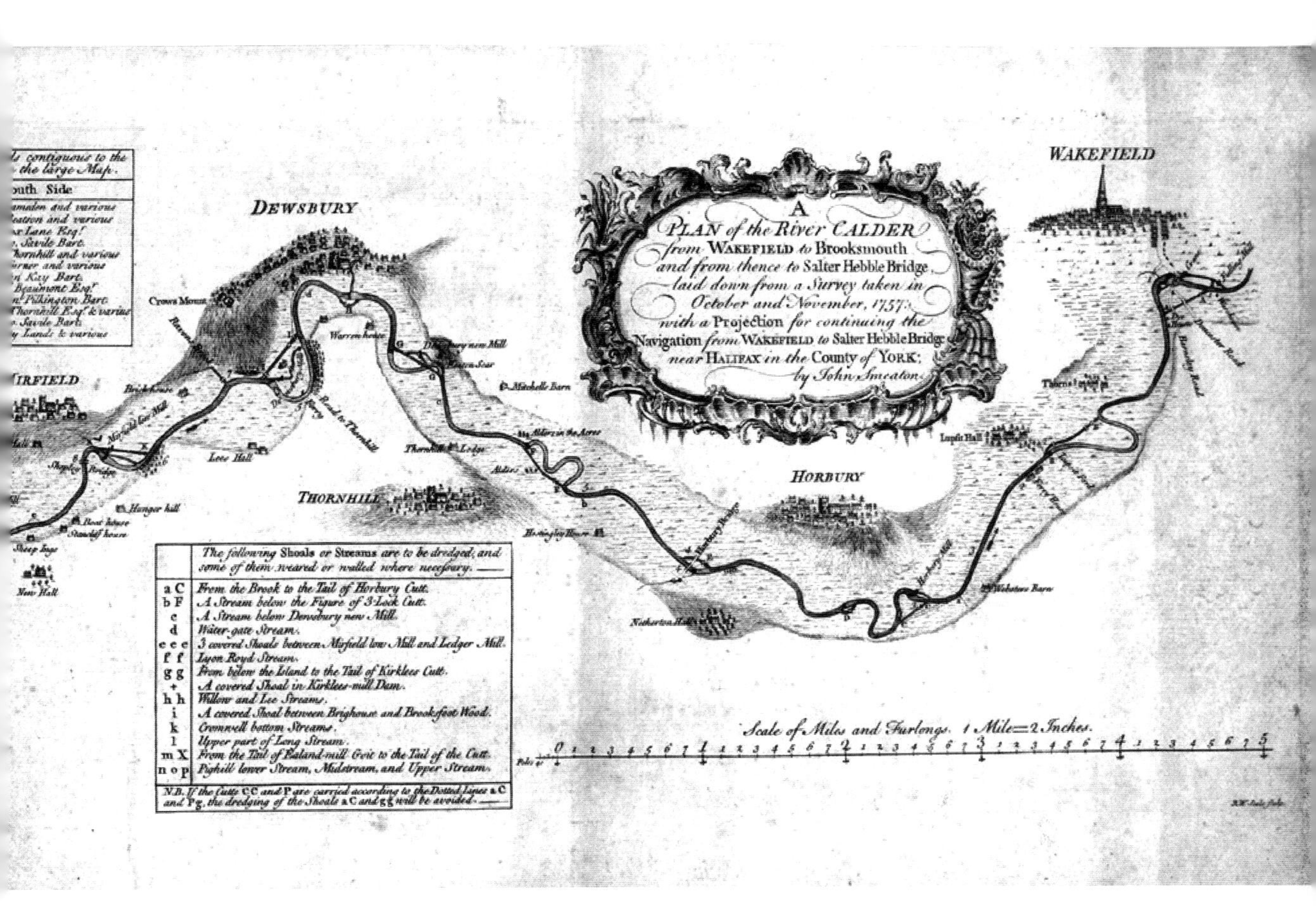
A
PLAN of the River CALDER
from WAKEFIELD to Brooksmouth
and from thence to Salter Hebble Bridge,
laid down from a Survey taken in
October and November, 1757:
with a Projection for continuing the
Navigation from WAKEFIELD to Salter Hebble Bridge
near HALIFAX in the County of YORK;
by John Smeaton.
WAKEFIELD
DEWSBURY
THORNHILL
HORBURY
Crows Mount
Warren house
Brick house
Lees Hall
Hunger hill
Boat house
Thornhill Lodge
Netherton Hall
Scale of Miles and Furlongs. 1 Mile=2 Inches.
The following Shoals or Streams are to be dredged; and some of them weared or walled where necessary.
a C From the Brook to the Tail of Horbury Cutt.
b F A Stream below the Figure of 3-Lock Cutt.
c A Stream below Dewsbury new Mill.
d Water-gate Stream.
e e e 3 covered Shoals between Mirfield low Mill and Ledger Mill.
f f Lyon Royd Stream.
g g From below the Island to the Tail of Kirklees Cutt.
+ A covered Shoal in Kirklees-mill Dam.
h h Willow and Lee Streams.
i A covered Shoal between Brighouse and Brooksfoot Wood.
k Cromwell bottom Streams.
l Upper part of Long Stream.
m X From the Tail of Ealand-mill Goit to the Tail of the Cutt.
n o p Pighill lower Stream, Midstream, and Upper Stream.
N.B. If the Cutts CC and P are carried according to the Dotted Lines a C and Pg, the dredging of the Shoals a C and gg will be avoided.

18. Cartouche from John Smeaton's Plan

and dealers in those towns. This was already a small victory for the navigation supporters as some of the latter communities had petitioned against the 1740/1 bill: even some of the mill-owners along the proposed navigation had been persuaded that it could be to their advantage. More supporting petitions came from further afield: Grantham, Lincoln, and Newark were keen to take advantage of easier transport for their grain and other produce up into the Pennine districts.

Despite the Halifax opposition, the 'Rochdale gentlemen, merchants and tradesmen' also petitioned to have the navigation extended up to Sowerby Bridge 'which lies on the Great Turnpike Road which forms the Communication between the Eastern and Western part of this Island.'[60] Eyes produced a survey for them of the additional stretch from Brooksmouth to Sowerby Bridge. Another petition for this proposal came from the Mayor, Aldermen, merchants and other inhabitants of Kingston-upon-Hull. The Lancashire gentlemen were successful and on February 10th 1758 when the bill was being considered in Parliament this was added as an amendment.

Despite the efforts of the navigation committee, there were several petitions against the extension. The 'Gentlemen, Merchants, Cloth-dressers, Woolstaplers, Dyers and other

60 JHC Vol 28: 31 Geo II Rochdale Petition.

Dealers and Makers of Broad and Narrow Cloth, Shalloons and Stuffs in Wakefield' claimed the extended navigation would be prejudicial to their trade. They were supported by a petition from the towns of Birstall, Batley, Dewsbury, Heckmonwike, Hightown, Heaton, Littletown, Gomersal and Mirfield, and another petition from the woollen merchants, clothiers and dealers in Leeds. Sir Lionel Pilkington, not unexpectedly, also presented a petition with Edward Darby and John Greenwood, all owners of mills on the river, who claimed that their mills had been maintained at great expense and that the proposed navigation would damage their property. The Parliamentary Committee duly considered all the petitions, and Lord Downe was appointed to report back to the House of Commons.

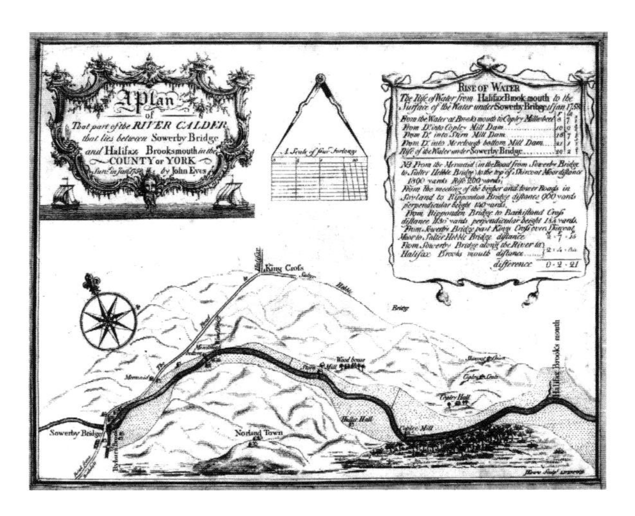

19. John Eyes' Plan of the River Calder between Brooksmouth and Sowerby Bridge, 1758

The exhaustive Parliamentary Committee hearings began with the Halifax petitioners. Their witnesses, David Stansfeld, merchant, and Samuel Lees, manufacturer and finisher of stuffs and cloths, outlined the advantages the extended navigation would bring to the upper Calder region, rehearsing the familiar arguments: it was an area with poor land and a very large population, with woollen manufacture being the main employment; wool came from the Eastern counties to serve this manufacture; corn and other provisions were transported to Wakefield or Leeds by boat, and thence to Halifax by road transport. They claimed that carriage by land was still expensive, indeed it had become more so since turnpike roads were established. Goods in Halifax cost considerably more than in Leeds or Wakefield, causing distress to the people at certain times of the year. If goods were brought by boat all the way to Halifax it would dispense with the need for middlemen, who also added to the costs.

Stansfeld was questioned further about the effect of the new turnpike roads, the introduction of which seemed to have had such a major influence on the failure to obtain a navigation in 1740/1. The road from Leeds had been completed but the price of carriage had not been reduced. The road from Wakefield was not yet complete, and pack horses or reduced wagon loads were necessary for this route. The final section, known as Halifax Bank, was 'too steep to be made good for carriages'. Stansfeld pointed out that other heavy goods could be carried by water: the export of flagstones from the area's quarries and the import of large quantities of lime needed in this upland region to improve the soil would be made much cheaper. With a navigation the roads would be used less for such heavy goods, and therefore repair costs would be far less.

In Smeaton's evidence to the House of Commons Committee he stated that he had surveyed the River Calder from Wakefield to Salter Hebble Bridge 'with utmost care'. Initially he presented the contents of his report estimating that the work would take seven years to construct depending on the number of hands employed. With the proposed depth of four feet it would be able to accommodate boats carrying twenty to twenty-five tons and a reservoir at Salter Hebble should supply sufficient water in dry seasons. In cross-examination he admitted that he had not been concerned in a river navigation before, although he had previously surveyed the Clyde in order to make it navigable. He made the point clearly that the great flow of the river would mean that the work would be expensive, but that this gradient would allow water to escape quickly.

The next petition considered was the Rochdale demand for an extension to Sowerby Bridge. George Stansfeld, a white kersey-maker from Sowerby Bridge gave evidence for this. The supplies of wool and corn to Rochdale also came from the eastern counties via Wakefield or Leeds. Equally their manufactured cloth was exported via Hull, being bailed up at Wakefield or Leeds. The long distances by packhorse over hard terrain sometimes caused damage to the goods and the lack of horses for this trade often led to long hold-ups in the movement of goods. The waterway would provide access to better, cheaper transport for many Pennine communities. The extension to Sowerby Bridge would mean that it was possible for pack-horses to travel there and back from Rochdale in one day. Stansfeld was followed by John Eyes who described his survey from Brooksmouth to Sowerby Bridge and gave his estimate of the cost for this stretch as £3,400.

The millowners' case was then explored. Nathaniel Bertram again represented these interests as the steward of Sir Lionel Pilkington's four corn mills, two fulling mills, and one frizzing mill at Wakefield. Bertram claimed that the navigation would lead to a loss of water

REFERENCES to the *WORKS* for *NAVIGATION.*

Letters of Reference	Places of the Locks &c.	Rise at the Locks	Length of the Cutts	Height of the Dams	Number of Bridges.
		Feet. Inch	Furl. Poles	Feet. Inch	Stone, Woo.
A A	*Fall Ing Lock*	10 . 5	2 . 27		2 . 1
B B	*Lupsit*	6 . 0	3 . 26	4 . 3	2
C C	*Horbury Mill*	9 . 3	2 . 3		1
D	*Washingstone Ford*	4 . 3	0 . 30	4 . 3	1
E	*Horbury Bridge*	6 . 1	1 . 20	5 . 0	1 . 1
F F	*The Figure of 3*	7 . 0	3 . 14	4 . 0	
G G	*Dewsbury new Mill*	8 . 6	2 . 0		1
H	*Dewsbury low Ford*	1 . 10	1 . 0	1 . 6	1
I I	*Dewsbury upper Mills*	12 . 3	3 . 2		2
K K	*Mirfield low Mill*	8 . 3	2 . 20		1
L L	*Ledger Mill*	8 . 6	1 . 26		1
M M	*Batty's Mill*	3 . 9	1 . 24		1
N	*Halls Mill*	5 . 0	1 . 10		
O	*Coopers Bridge*	6 . 3	1 . 6	6 . 0	1
P	*Kirklees Mill*	7 . 0	1 . 32		1
Q	*Anchor Pitt*	4 . 8	1 . 17	3 . 6	1
R R	*Brighouse*	10 . 8	3 . 14		1 . 1
S	*Lillands*	5 . 0	0 . 37	4 . 0	1
T	*Taghole*	6 . 6	1 . 8	5 . 6	
U U	*Gool*	3 . 10	1 . 14	2 . 6	1
W	*Ealand park Nook*	6 . 0	1 . 23	4 . 2	1
X Y	*Ealand*	6 . 6	3 . 0		1
Y Y	*Long Lee*	6 . 6	1 . 0	6 . 6	
ZZZ	*3 Brook Locks*	24 . 0	2 . 30	*and a Reservour,*	1
No. 26	from Wakefield to Salter Hebble Brid.	178 . 0	M. F. P. 5 . 6 . 23		12 . 12

20. Panel listing mills, locks and dams from John Smeaton's Plan

when water was already often scarce, through the leakage of the lock gates and boatmen who 'flashed'[61] their boats through the locks. He again described the nature of the Soke which meant that the Soke's corn mills required constant maintenance. Fulling sometimes had to stop to make sure the corn mills kept going, although in summer all mills might stop for lack of water. In winter either backwater or low water could stop the working of the fulling mills and spoil any cloth as it was being fulled, a process which took from ten to twenty hours. The mill below the bridge at Wakefield was subject to this problem because of a dam a mile downriver. At Knottingley and Castleford too the mills were often affected by backwater caused by weirs or loss of water and the Aire and Calder Navigation recompensed them, paying £200 a year. There were, he calculated, 30 fulling stocks[62] between Wakefield and Brighouse which would be affected by the navigation. Other witnesses described the problem of backwater which could halt the fulling process and cause damage to the cloth. The final witness against the navigation, Joseph Willis, noted the falling cost of land carriage which was his preference, especially with the greater use of broad-wheeled wagons or stage wagons which were better for the roads. Developments in road transport arrangements meant that regular 'back carriages', one-day return journeys, were increasingly efficient.

Smeaton returned to reply in detail to these concerns, explaining that there would be no need to flash through locks because the locks would be constructed 'as they ought to be'. He mentioned his work on the Soke Mill in 1754 when he conducted an experiment to find the quantity of water sufficient to keep the wheels turning. His familiarity with the Calder previously was useful as he had experience of changes in river levels. Leakages of the locks on the Aire and Calder were great because the gates were not properly made, whereas some of the locks he had observed in the King's docks and abroad did not appear to have any leakage. He would ensure that gates on the new navigation would only leak to a small degree. In addition he could advise about improvements to the mills on the Calder so that they would work with greater efficiency using a third less water. Backwater was far less likely on a fast-flowing river than in flat country and his dams would not raise water high enough to become stagnant. He had also calculated that twelve new sets of mills could be built on the dams along the new navigation. His cross-examiners asked about sealing the locks and the foundations for buildings. The proximity of stone to build securely was emphasised. He mentioned work on the Eddystone where he had employed methods of setting stones in water. His evidence was backed up by George Collett who

---

61 In cases where there was only a single gate a boat would wait above the lock, the gate would be raised and the boat 'flashed' through with the flow of water. Going upstream the boat would be hauled through by men or sometimes winched through. This system was used where there were mill weirs and Bertram was concerned about such a weir below Horbury Mill, which was part of Wakefield Soke.

62 Fulling stocks were the water-driven hammers which pounded the woollen cloth in order to eliminate oils, dirt, and other impurities, and make it thicker.

was Master House-carpenter to the king. He had designed the lock gates at Deptford - the leakage of which was 'trifling' - and he answered questions about keeping gates in repair and the strength of piling to secure the gates.

Questions about fulling were also put to Jonas Crowder a fulling miller at Sowerby Bridge, and David Stansfeld (again examined) who had a fulling mill on the Calder. They were asked to describe in some detail how stoppages affected the fulling process and in particular if it was possible to treat cloth which had suffered stoppages of limited time periods. The fulling process, he argued, was not usually irretrievable if cloth was not left in the stock. Stansfeld felt that more fulling mills could be constructed outside the intended navigation and the present fulling mills were not usually used to the full. The navigation could be an advantage to the millers in getting fullers' earth more cheaply as well as corn, timber, stone and lime. Stansfeld's response to the evidence that road transport costs had fallen and more goods were being carried in stage wagons was that there had been no difference to prices on the Leeds to Halifax road. Stage wagons between Halifax and Wakefield were not useful because of the lack of control on the steep hills. During his evidence he stated that the Aire and Calder Navigation was more costly than it needed to be: the proprietors he said took 30% or 40% of the profits on lock duties.

After this detailed examination of the evidence it was ordered that leave should be given to bring in the necessary Bill, and Lord Downe, Sir John Armitage, Mr Hewett, Sir Conyers D'Arcy, the Lord Strange and Mr Bold were asked to prepare and bring in the same. By 10th April the Bill had been read for a second time, but on the 19th April five new petitions were sent from Bradford, from Mirfield and other villages nearby, from villages round Wakefield and Dewsbury, Birstall and Batley, from Horbury, Emley and other villages near Huddersfield and from the Mayor, Aldermen and inhabitants of Beverley questioning the powers of the undertakers over the use of the tolls and profits. As it stood the bill stated that those who invested in the new navigation would be undertakers for carrying on and continuing the navigation with power to take tolls upon goods and vessels as their own property. The new petitions were asking that the tolls and profits should be vested in trustees, not undertakers who kept the tolls high for their own profit, rather than for the benefit of the users of the navigation. After the trustees had paid off what had been borrowed for construction they could lower the tolls retaining only sufficient money to cover repairs and other works. Eventually amendments to this effect were included in the bill.

The Act received the Royal Assent on 9th June1758.[63] Over eight hundred men from the West Riding and Lancashire, listed in order of social standing and all qualified by substantial property ownership, were appointed commissioners. A quorum of no fewer than nine were empowered to appoint agents and workmen, and be responsible for the

63 JHL, Vol 29: June 1758 Act of Parliament: 31 Geo II c.72.

construction of the navigation, deepening and straightening the river, making new cuts and dams, raising banks and removing obstructions like trees. They were also responsible for making new bridges over the cuts and replacing fords, for constructing sluices, locks, weirs, warehouses, winches, engines and weighing beams. They were empowered to repair and enlarge ways to the waterway, alter bridges, and make towing paths. They could not however alter access to houses, orchards or paddocks and were required to recompense damage to property along the river. Most importantly they were entitled to buy land, and other property they required for the work. Where an offer had not been accepted within thirty days they had the power to apply to the Sheriff for a jury of twenty-four to make a binding judgement. The commissioners were to appoint a collector of tolls, a treasurer, and other officers who would be salaried.

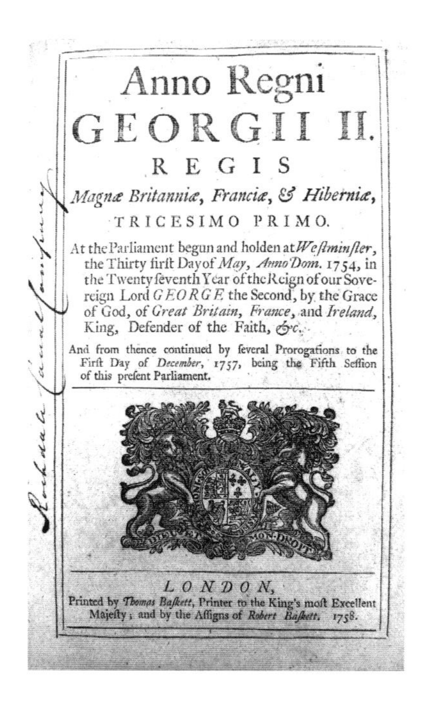

Anno Regni
GEORGII II.
REGIS
Magnæ Britanniæ, Franciæ, & Hiberniæ,
TRICESIMO PRIMO.

At the Parliament begun and holden at *Weſtminſter*, the Thirty firſt Day of *May*, *Anno Dom.* 1754, in the Twenty ſeventh Year of the Reign of our Sovereign Lord *GEORGE* the Second, by the Grace of God, of *Great Britain*, *France*, and *Ireland*, King, Defender of the Faith, *&c.*

And from thence continued by ſeveral Prorogations to the Firſt Day of *December*, 1757, being the Fifth Seſſion of this preſent Parliament.

LONDON,
Printed by *Thomas Baſkett*, Printer to the King's moſt Excellent Majeſty; and by the Aſſigns of *Robert Baſkett*. 1758.

21. First page of printed Act of Parliament 1758: 31 Geo II c.72

Rules for the usage of the waterway followed, with penalties for breaking them. The main heavy goods were to be charged for by the ton: stone, slate or flags, lime and limestone, coal were specified, with exemptions for stone, timber, gravel sand and other goods for the millowners. Similar tolls by the ton were charged for the carriage of 'goods, wares, merchandise and general commodities', not further itemised. Of particular note is the clause dealing with the transport of coal. Presumably inserted to protect the supply and pricing of coal to the upper Calder communities, the carrying of coal cargoes downstream towards Wakefield was to have a penalty of £50. This deliberate limitation on the use of the waterway for coal suppliers from this part of the Calder Valley would certainly have been unpopular. It may also have been difficult to enforce, and it was not included in the later Acts.

CHAPTER 2

AFTER THE PASSING OF THE ACT

The Upper Calder Navigation Act became law on 9th June 1758[64] but John Smeaton's Journal, transcribed in this volume, did not begin until 26th May 1760. The period up to June 1758 had been intense and exhilarating for the supporters of the navigation, but what followed was a period of very intermittent activity which taxed the relationship between the gentlemen in Halifax and their superintendent engineer.

At the Market Crosses of Halifax, Wakefield and Rochdale and in the local papers, notices announced the first meetings of the navigation commissioners. Although the Act had named a huge list of men from the region, qualified by substantial property ownership to act as commissioners, only those with the most interest in the progress of the scheme attended. At their first meeting on 7th July at the house of Mr John Mellin, the Talbot in Halifax, the newly-appointed clerk, John Bentley, was empowered to write to John Smeaton, then in Plymouth, expressing their resolution that he was 'the properest person to superintend and construct the Works', to ask when he would be available, and what his terms were.[65] In the following meeting on 27th July those who were prepared to take on the role of a commissioner were sworn in. The Oath was taken by forty-one men, many of whom had already been active in the work leading to the Act. They promised that they would 'without Favour or Affection, Hatred or Malice, truly and impartially, according to the best of their Skill and Knowledge, execute and perform all and every the Powers and Authorities as a Commissioner' in pursuance of the Calder Navigation Act. Sir George Savile was again at the head of this list.[66]

---

64 JHC Vol 28: 31 Geo II.

65 WYJS/CA MIC 2/4.

66 WYJS/CA MIC 2/2.

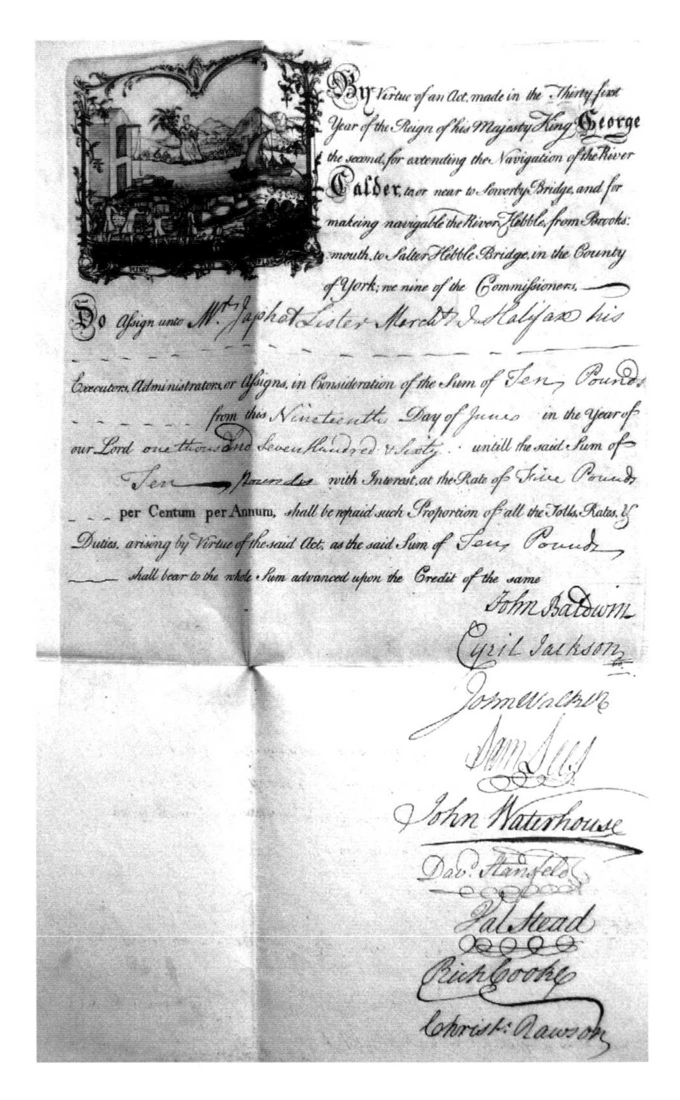

22. Subscription of Japhet Lister, Merchant of Halifax of £10, 19 June 1760. with Transcription

By Virtue of an Act made in the Thirty first Year of the Reign of his Majesty King George the second, for extending the Navigation of the River Calder, to or near to Sowerby Bridge and for makeing navigable the River Hebble from Brooksmouth, to Salter Hebble Bridge, in the County of York; we nine of the Commissioners,

Do Assign unto Mr Japhet Lister Mercht. in Halifax his Executors, Administrators, or Assigns, in Consideration of the Sum of Ten Pounds from this Nineteenth Day of June in the year of our Lord one thousand and Seven Hundred & Sixty, untill the said Sum of Ten Pounds with interest, at the rate of Five Pounds per Centum per Annum, shall be repaid such Proportion of all the Tolls, Rates & Duties, arising by Virtue of the said Act, as the said Sum of Ten Pounds shall bear to the whole Sum advanced upon the Credit of the same

John Baldwin, Cyril Jackson, John Walker, Sam Lees, John Waterhouse, Davd Stansfeld, Val Stead, Rich Cooke, Christr Rawson

The first job of these gentlemen was to organise the borrowing of sufficient money to pay for the work. Subscribers to the navigation would pay instalments of a promised amount, and would receive yearly interest on these paid subscriptions of not more that 5% until the navigation itself could generate income from 'tolls, rates and duties' to repay them. Work was not to begin, they resolved, until £36,000 had been subscribed. Potential subscribers were approached and large amounts were promised. For example on 26th August 1758 R Yeates felt the pressure to appear generous after he had seen a list of subscriptions: 'I observe that so small a Sum has not been subscribed and I should be very sorry to set a bad Example. And as I find that the appearance of my Name upon this Occasion will be agreeable to yourself and the rest of the Gentlemen you may subscribe in my Name Three Hundred Pounds ... '[67]

Other subscriptions promised on 29th August 1758 were £500 from the Rev Hargreaves, Curate of Todmorden,[68] and £200 from Elin Barber.[69] Later, in 1764 George Stansfeld who had organised these subscriptions felt able to write to Hargreaves: 'The Navigation succeeds surprisingly, most likely you will see Boats pass next year under your estate to Sowerby Bridge'[70] In the event his prediction was premature. At the third meeting of the commissioners on 31st August 1758 Mr John Caygill and Messrs Jeremiah Royds and Co, were appointed as Joint Treasurers. The Leeds Intelligencer reported that already above £30,000 was subscribed by 'forty Gentlemen and Merchants'.[71]

By 12th September the first subscription payments were requested at 'six pounds per Cent, upon their respective Subscriptions.'[72] The rapid response did not appear to have been affected by the amendments to the financing of the project late in the passage of the bill. The proposers of this amendment had been regarded at the time as enemies, risking potential investment into the scheme. However, now more assured of the finance, the commissioners were proud of their own public-spiritedness, doubtless in comparison with the dividend-earning but much criticised Aire and Calder Navigation undertakers. They announced in the newspaper: '... there will not be wanting men of virtue and publick spirit to carry into execution such a large and extensive work as this, when solely calculated for the Benefit of the Publick in general.' Evidently Sir George Savile had set the example for this when he had refused to become an undertaker under the Bill's previous arrangement, 'but when vested in Commissioners with powers to borrow money at common Interest, and stript of all future personal advantages, he most generously and cheerfully gave his assistance'.[73]

---

67 WYAS CA FH: 374.
68 WYAS CA FH: 375/2.
69 WYAS CA FH: 375/1.
70 WYAS CA FH: 375/4.
71 LI 12th Sept 1758.
72 LI 19th Sept 1758.
73 LI 12th Sept 1758.

At the meeting of 31st August 1758 the gentlemen listened to Smeaton's carefully worded letter sent from Plymouth in reply to Bentley's question.[74] Whilst couched in the most polite terms it laid out with great clarity what he considered his role in the work should be. He should not be expected to be constantly in attendance upon the works but would do the 'designing, planning and laying out'. He assured them that 'I am too solicitous for the success of whatever I undertake to trust the execution of any new improvements or contrivances, or anything where there is more than ordinary difficulty to any other than my own eyes.' but that a Deputy Surveyor should be appointed 'to be constantly upon the works' at a suggested salary of £100, and it would be his job to see Smeaton's plans 'put in strict execution without deviation'. After reiterating these terms 'with other such service as the Commissioners shall require of me, or shall fall out within my province' he again stressed his superintending role.'I hope it will not be thought unreasonable that I should be at perfect Liberty to undertake any other Business that shall not be inconsistent with those engagements.' He spelt out the areas he would not expect to be involved with at all: accounts or any money transactions, making contracts or 'measuring any work done thereby', although he might advise if required. He offered a choice of terms for his remuneration: 5% of the costs of the works, £300 for the first two years and £200 after that, or £250 a year.

Finally he discussed his present commitment to finish the work upon which he was engaged at the Eddystone Lighthouse, which was highly dependent on the weather. He expected that he might leave there at the end of October or early November in 1758, but would need to return the following year, for at least some part of the summer. He then began to offer advice about tools and materials for the Calder work and promised to furnish the meeting after the next with 'a model or design of the general method of carrying on the Works' as well as a recommendation for a Deputy Surveyor. The commissioners considered the letter and Smeaton's terms were accepted, 'under the limitations you desire' suggesting some disappointment. They 'asked that you give them leave to hope ... you will give your Superintending Care in such contracts and other matters as may exceed the skill of your deputy or where for any other reason it may be thought necessary by the Commissioners.'[75] He was to be paid £250 a year and to have the choice of his deputy surveyor. It is difficult to gauge the commissioners' response to the business-like presentation of Smeaton's intended approach to the work and the timetable he imposed: he would not be available for extended employment with them until the end of 1759 and, even then, he would not oversee the works on a full-time basis. After the excitement at the passage of the Act there must have been a sense of anticlimax felt by some of the commissioners.

---

74 WYJS/CA MIC 2/4 15th Aug 1758.
75 WYJS/CA MIC 2/4 15th Aug 1758.

The next meeting was set for November 15th when the commissioners 'hoped to have the Pleasure of seeing' Smeaton.[76] During the September and October of 1758 an alternative method of raising shipping by crane rather than building locks was offered by John Kemmet and Co. who had patented a machine for this purpose and proposed favourable terms for completion of the work. The commissioners were interested to find out about it, although it had not been used on any river, and Smeaton's opinion was sought as the overall decision-maker. His response was to say he would waive any engagements for the work if they wanted to employ others. They invited Kemmet to a meeting to explain his machine but he eventually declined and on 15th November 1758, this offer was formally rejected. Smeaton had been expected at this meeting but was not in Yorkshire. Evidence of their frustration is to be seen perhaps in a resolution made to employ a Master Carpenter and a Master Mason to serve jointly as Deputy Surveyor, thus challenging the agreement over Smeaton's terms. However, they did approve Smeaton's purchase of pozzolana and other tackle from the Eddystone. At this meeting the sum spent on the application for the Act was announced at over £2,000, an amount that must have brought home the scale of the whole commitment being undertaken.

Their final meeting of the year was on 29th November but again Smeaton had not yet arrived in Yorkshire. Because of Smeaton's various commitments and the time-consuming and exhausting nature of travelling, particularly in winter, his journeys had to be planned with care. A one-way journey from London to Leeds at this time took at least three days: the '*Flying Machine on Steel Springs*' travelling from Leeds to London via Wakefield, Barnsley and Sheffield was being advertised in 1760 as offering a three day journey,[77] or a gentleman might hire a post chaise which could carry two people.[78] However, Sir George Savile attended and the potential for dissent and confusion arising from the unwieldy commissioners' meetings was recognised. A committee was chosen from the gentlemen, consisting of five members, two of whom could act on behalf of the commissioners until the Annual Meeting in June, allowing communications and decision-making to become more consistent and speedy. Those chosen until the end of June were William Gream, Mr Stead, Dr Jackson, David Stansfeld and Mr Baldwin. The commissioners continued to meet regularly and the committee members were able to report on developments such as their progress in sourcing timber and other correspondence.

Meanwhile, as Smeaton had anticipated, work on the Eddystone stopped for the season in early October 1758, but he then visited a number of other possible projects. In December 1758 he was on the Clyde to prepare a report on the Great Canal and during the trip he consulted on the Seaton and Cullercoats Harbours. During this trip north he

76 WYJS/CA MIC 2/4 31st Aug 1758.

77 LI 17th June 1760.

78 LI 6th Jan 1761 Advertisement for a travelling companion in a Post chaise.

23. Photograph of Austhorpe Lodge (undated)

was asked to advise on the possibility of a navigation on the River Wear too.[79] Eventually he arrived at Austhorpe Lodge, Whitkirk, near Leeds, his much-neglected family property, but even then he did not manage to attend any commissioners' meetings. On the date of their meeting of January 15th 1759, he was indisposed and sent his apologies for not travelling to Halifax. However in the same letter it is clear that he had been very active in moving the project forward. He reported on wood he had inspected and on his review of suitable quarries along the river. He had already invited Joseph Nickalls, millwright from Lambeth, to be his deputy, recommending him to the commissioners as a 'person tho' a common working millwright, and the greatest part of his practice has been in that way, yet I have seen some difficult sluice work executed by him, in so masterly a manner that I thought I should be very happy in having such a person to putting designs in execution'. Smeaton suggested to the meeting that Nickalls should be employed soon 'to work with a few hands in constructing the several Engines for water and piles etc that will be wanted in building work, houses, forges etc and getting every convenience ready with the necessary materials for an actual commencement of the work at the beginning of April twelve months'. He told the commissioners that he had been conducting trials to perfect a lock which would not leak and that he had already begun to discuss the new machines he intended to use on the Calder with Nickalls. Thus at this commissioners' meeting of 31st January they were given the date at which Smeaton anticipated the building of the Navigation would start. They confirmed Nickall's appointment to begin at Michaelmas 1759 [80], and would provide £21 for his removal costs: Smeaton was also asked to find a master mason at not over one guinea a week. They agreed to Smeaton's view that the first stretch of the Navigation should be between Dewsbury and Wakefield.

On his return to London in February 1759 Smeaton gave evidence to Parliamentary Committees for the Wear and Clyde projects, and in March and April 1759 for a Wey Navigation. However, uppermost on his mind in this early spring must have been the completion of his paper for the Royal Society which he returned to London from Plymouth

79 Skempton, Ch 1 'John Smeaton' Trevor Turner and A W Skempton p 15.
80 WYJS/CA MIC 2/4 29th Sept 1759.

to read over five evenings in May and June 1759. His work on the Eddystone Lighthouse itself began again in early July 1759 and proved more demanding than expected as a 'distemper' had decimated Smeaton's workforce. Eventually he arrived back in Plymouth from the Eddystone on 9th October in 1759 but remained in the West Country to wind up the whole business. On his return to London he prepared a plan for the Blackfriars Bridge, although he did not win this commission. His final appointment in London on 22nd November 1759 was to dine at the Royal Society to receive the Copley Medal awarded for the paper he had read earlier in the year.

24. William Jessop about 1797

During all this period Smeaton's personal circumstances had been changing. Smeaton's widowed mother moved to London after her husband's death in 1749, and she died there in October 1759. Smeaton had married Ann Jenkinson of York in 1756, and in 1757 the couple lost their first child, Hannah. In 1759 a second daughter, Ann, was baptised. Although the family were based in London, Smeaton inherited the property at Austhorpe near Leeds where he was able to stay whilst in Yorkshire.[81] As the Calder work began he took a fourteen-year-old apprentice, William Jessop, whose father had been in charge of the Eddystone operations: the boy started his apprenticeship at Austhorpe copying Smeaton's plans.[82] Jessop was later to become a leading canal, harbour and railway engineer himself.

Smeaton had maintained contact with the clerk to the commissioners by means of detailed letters, but by the early autumn of 1759 an increasing sense of frustration amongst some commissioners seems to have developed. He was never available in Yorkshire to attend their meetings and the beginning of the work still seemed distant, with little achieved. The experiments, the scientific groundwork, the status he was earning from his reports and appearances in Parliament would have been largely unseen by many of the gentlemen in Halifax and Rochdale as another autumn found them waiting for him to come to Yorkshire. Smeaton wrote a long letter from Plymouth on 16th October 1759 in which he referred to issues raised in correspondence from Doctor Jackson. There was clearly a question in Halifax over his commitment to the project. He patiently explained

81 WYJS/CA MIC 1/4 15th Jan 1759, 19th Dec 1759.

82 http://www.engineering-timelines.com/who/Smeaton_J/smeatonJohn12.asp.

the delays at the Eddystone, assuring them that 'it is my resolution punctually to perform my engagement'.

The main dissatisfaction seemed to be the rumours that he was taking other work, not only the plan for Blackfriars Bridge, but also an offer of work in Scotland. Smeaton should give all his time to the Calder, they thought, and then another Engineer would not be necessary. For an unknown reason, perhaps connected to this view, Nickalls had not started work at Michaelmas as previously agreed. Smeaton was forced to restate his position, referring to his first letter to the commissioners. They had accepted his terms that 'whenever my attendance upon the Works on the Calder as Engineer in Chief was not Necessary that my own time should be at my disposal and that with this in View a skilful person should be entertained as Deputy Surveyor to take charge thereof in my Absence.' He continued 'Some few I understand Oppose it (Numbers are always attended with differences of Opinion)'.

His acceptance of other projects was important to him at the time: he needed a comfortable home for his growing household and a good workshop and study facilities for his business: Austhorpe Lodge required substantial work to provide this. He charged consultancy fees of £1 a day with travel expenses but his income still remained limited, and so he needed to accept work other than his main commitment on the Calder. Additionally his work at the Eddystone had been physically demanding and he was learning to protect himself from such heavy demands again.

By 9th November he was completing his affairs in Plymouth and travelled with his family to London. On the 28th November he was at the commissioners' meeting in Halifax to begin his employment with them. A description of his duties was on the agenda:

> 'That his Duty is understood to comprise in the following particulars:
>
> That he take his instructions from the Commissioners and in Consequence to take all original Levels and Dimensions in Order to make out Plans and Designs for the approbation of the Commissioners and Instruction of the Workmen not only of the several Cuts, Locks, Dams, Bridges and other Works to remain upon the River, but also Designs for the several Engines Machines and Utensils for facilitation the Execution thereof to lay down or mark out the several Designs upon the Ground to attend the execution of all new or difficult Parts of the Work and at such other Times also as nine or more of the Comms shall judge necessary and give him Notice to do, and to see that every Office under him does his Duty according to his Instructions and in Case of Non-performance to report the same to the Commissioners But when Time or Circumstance will not permit that to be done, to have power to discharge such Officer or act therein according to Discretion and
>
> To keep a Journal giving an Account of the State and progress of the Works as shall appear from Time to Time upon his Inspection and to report the same to the Commissioners.

To Make Estimates of the Quantity, Quality of the Materials necessary for each Work and when to be provided at all other Times when his Attendance is not necessary in such Respects as above specified to be at full Liberty to imploy himself in what ever other undertakings he should think proper.

All Salaries were to be paid quarterly: Christmas, Lady Day, Midsummer and Michaelmas.'[83]

During early December Smeaton stayed in Wakefield, probably at the White Hart, to inspect the sites for the first three cuts and continue his negotiations to obtain stone. On December 13th 1759 he informed the commissioners that he needed to retire to Austhorpe 'in order to be free from the Interruptions that the living in a public house and in a Town where I have so many acquaintance naturally subject me' and to prepare estimates for the three cuts. His plans needed adjustments to be ready for the commissioners to view and for Nickalls to use. Joseph Nickalls finally arrived at Austhorpe on Christmas Day and on the 26th of December the two men returned to Wakefield to look at the quarries and then attend the commissioners' meeting on 2nd January. At this meeting Joshua Wilson was appointed Master Mason, and it was agreed that the posts of a foreman smith and a foreman carpenter were to be advertised. As before, the immediate programme of works for Smeaton and Nickalls was precisely listed at the meeting: everyone knew what needed to be done.

'Ordered the Mr Smeaton and Mr Nickall shall contract for such Stone, Timber, Iron, and other Materials proper for carrying on the Works

Also that they shall mark out the Grounds for the Cutts and other Works at Horbury Mill, those from Thorns to Lupsit, and those at Wakefield according to the Plans deliver'd in by Mr Smeaton and to contract with proper persons for digging the same and to mark out such Grounds as may be necessary for laying Materials, building Workhouses and such other uses as may be necessary for carrying on the Works of the Navigation

Also to erect such Work shades and construct such Engines, Machines and Vessels and other Utensils as shall be judged necessary by Mr Smeaton or Mr Nickalls for putting said Works in execution, and to employ or discharge Workmen, Labourers and other Agents and Servants for the Construction thereof and of the Works themselves as they shall judge necessary.

That the three Cutts from Fall Ing to Horbury Mill Dam from the plan of which Mr Smeaton has laid before the Commisioners be made so wide as that two boats can (pass) and repass each other

83 WYJS/CA MIC 2/2 28th Nov 1759.

That Mr Smeaton and Mr Nickall stake out the Ground first for the Width of the Cutts at the Top and 2ndly for Banks on each side.

That all stone be paid for when delivered and that all Materials to be bought be paid for upon Delivery or as otherways agreed for.

That the mark CN be put upon all markable Materials.'

The two men then stayed in Wakefield working on the marking out. Smeaton reported progress by letter in detail to the commissioners every two or three days. By 11th January 1760 he announced that the project was ready for the purchases of land as the first three cuts at Wakefield, Lupset, and Horbury had been staked out and he had measured the land required from each proprietor. The plans of the projected cuts were sent to the commissioners with the quantities of materials required. With Nickalls and Wilson he then 'resurveyed the principal quarries and the present locks on the Calder' and decided on the location of the works yard beside the projected cut at Wakefield.

It seems that Smeaton and Nickalls left Wakefield together soon after this, calling at York to look at some timber from whence Nickalls went to Hull to buy wood, and Smeaton to London. At a meeting of the Commissioners on 3rd October 1759, Thomas Simpson had been appointed as pay clerk, and clerk to the commissioners at a salary of £50. From 1760 the commissioners' letter book is dense with correspondence between Simpson and Smeaton, Nickalls, the suppliers of materials and labour, and the land owners, revealing the great ferment of activity as the project swung into operation. At the same time Nickalls and Smeaton were constantly exchanging letters so that the one was always informed of developments and the other able to speak for his chief engineer on all issues. In the event Nickalls, the clerk and committee were able to organise the final preparations without Smeaton being in attendance.

As usual during his London stay Smeaton experimented, this time on Barrow-on-Soar lime,[84] and finding it equal in quality to the Watchet lime he had used at the Eddystone he arranged for a cargo of it to be sent to the Calder works with the committee's approval. He was also drawing up his final plans for the locks: on 17th April he wrote to Simpson asking for a decision about the width of the locks. He laid out the possibilities: during the preparation of the bill he had used the dimensions on the Aire and Calder which admitted vessels of 13'6" wide but he had noted that the locks below Knottingley were 17'6" wide to admit the Knottingley Keels. He recommended that the new locks should be wider than 13'6' to avoid damage to the walls and gates but wondered if they should be equal to those below Knottingley in case the locks below Wakefield were widened at a later date. However this would add at least £500 to costs and require much more water for the passage of boats. By 28th April the reply came that the locks should be 14'6" wide.

84 WYJS/CA MIC 2/4 22nd April 1760.

Another of his tasks in London was to visit the gentlemen who were up in town who owned land required for the first three cuts.

Finally Smeaton reappeared in Yorkshire:

> 'This Day I propose setting out with my family for Yorkshire, and tho I am here at a more advance Season than I expected yet as I expected to be down as soon as the Diggers can be ready, and have kept a punctual and regular Correspondence with Mr Nickalls concerning the Works I hope the Committee are also sensible that no real delay has been occasioned to the Works on this account..'[85]

The Calder Navigation was his major commitment for some years, and he brought his household to Austhorpe for the season although he retained his London home until 1763 when he was able to move his apparatus into his new workshop in Yorkshire. He continued to spend the early months of every year in the capital even after 1763, often giving evidence at Parliamentary Committees, as well as attending the Royal Society, and continuing to research, experiment, and write his reports and papers.

85 WYJS/CA MIC 2/4 22nd May 1760.

CHAPTER 3

PREPARATIONS AND PROGRESS

Letters and minutes in the Calder and Hebble Archive help to provide more detail about the period from the passing of the Act to the final entry of Smeaton's Journal in November 1763. They reveal the preparations, the progress and the problems faced by Smeaton and the commissioners. The location and collecting of the materials required for the work, the acquisition of the land, and the employment of the work force show the great scale of the activity which, for a short period, was focused on Wakefield waterfront.

Sourcing Stone

One of Smeaton's prime concerns was to locate quarries that could supply the type and quality of stone needed. Large blocks of good quality stone were needed for the sills and the hollow posts that held the lock gates. It could take time for the quarries to uncover these larger blocks, the quarries producing at first smaller stone suitable for 'backing' and for 'setters'.[86] Yorkshire sandstone differs in colour depending on the quarry, and Smeaton refers to 'bluestone' which has a bluish tint as opposed to others that are buff or cream; this was a better quality stone. There are also many references to the type of stone being dug out of the cuts. It is clearly an advantage if stone can be found close to the route, but to get the quality and size for certain parts of the masonry he was prepared to go further afield.

In 1758 Smeaton was already visiting quarries and assessing their suitability. When Joseph Nickalls was appointed, he and Smeaton were formally given responsibility for all contracts to do with stone, timber and other materials, and it was minuted that all stone should be paid for when delivered unless otherwise agreed.[87] On 2nd January 1760 Joshua Wilson was appointed Master Mason and shared in evaluating quarries.

86 Setters were rectangular blocks that were used on the lock pit bottom, and elsewhere for paving, also known as setts; backers were rubble masonry used behind the facing stones.

87 WYJS/CA MIC2/2 2nd Jan 1760.

The letters and reports demonstrate in some detail the progress of surveying the quarries and negotiating with owners and quarrymen. At the end of 1758 Smeaton wrote to the commissioners:

> 'With respect to the place of beginning the foot of the Navigation is undoubtedly the most natural provided the necessary materials etc can there be procured with advantage, and as the Tonnage of the stone will be more than four times greater than all the rest, this is the principal Article to be attended to; with this view I have examined All the principal Quarries about Wakefield that are likely to be of service'[88]

Smeaton stated that with the start of work at Wakefield: 'plenty of stone may be procured at a moderate distance fitt for filling and backing the walls but the only quarries producing good Aisler[89] is upon Heath Common and upon the Moor Adjoining New Millar dam ... There are also quarries of good hard stone near Horbury Bridge both on the Horbury and the Netherton side but it rises in such uncouth slopes and amongst such immense quantities of wast.' Smeaton decided to start with stone from Heath: ashlar at 'Wakefield Lock will be about 5000 feet besides setters backers etc which may be got out of a quarry in the Grounds of Mr Smith of Heath within a quarter of a mile of the river in case he can be prevailed upon to let it be worked; the Stones might from thence be brought up to Wakefield by water through a part of the old Navigation ...' [90] Although the quarry at Heath was close enough for stone to be transported by water, Smeaton was unsure how durable the stone would be under water.[91] The six-inch Ordnance Survey maps of the 1850s are an indication of where the quarries may have been sited, although much may have changed between Smeaton's time and the date of the maps. Quarries to the north and to the south of Heath Hall are shown.

Negotiating with the owners of the quarries could be frustrating. Smeaton hoped to obtain stone from Newmillerdam. These quarries, possibly the ones shown on the 1850s map to the east of the lake, and on the moor to the west, were owned by the Duke of Leeds who leased them to Sir Lionel Pilkington. However the right to work the stone was claimed by a Mr Bowes. Negotiations were entered into with the Duke of Leeds' agent, William Marsden, Smeaton eventually meeting him on 1st December 1759; however proceedings did not go smoothly. The Duke of Leeds was asking for a half penny a foot, and considering the distance that the stone would have to be transported, this was regarded by Smeaton as too much, particularly as 'the Gentlemen were in Hopes, that as this priviledge had hitherto been granted for what was next to Nothing that they shoud be indulged so

---

88 WYJS/CA MIC2/4 15th Nov 1758.

89 'Aisler' or ashlar - finely dressed, worked stone.

90 WYJS/CA MIC2/4 15th Nov 1758.

91 WYJS/CA MIC2/4 15th Nov 1758.

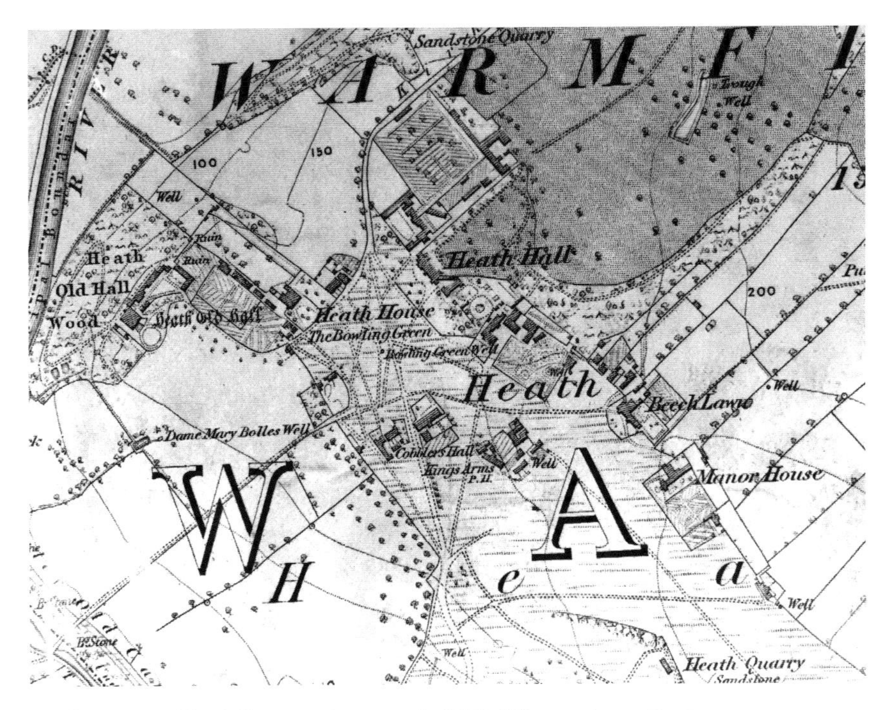

25. *Quarries on Heath Common shown on the 1850s OS map, close to the river*

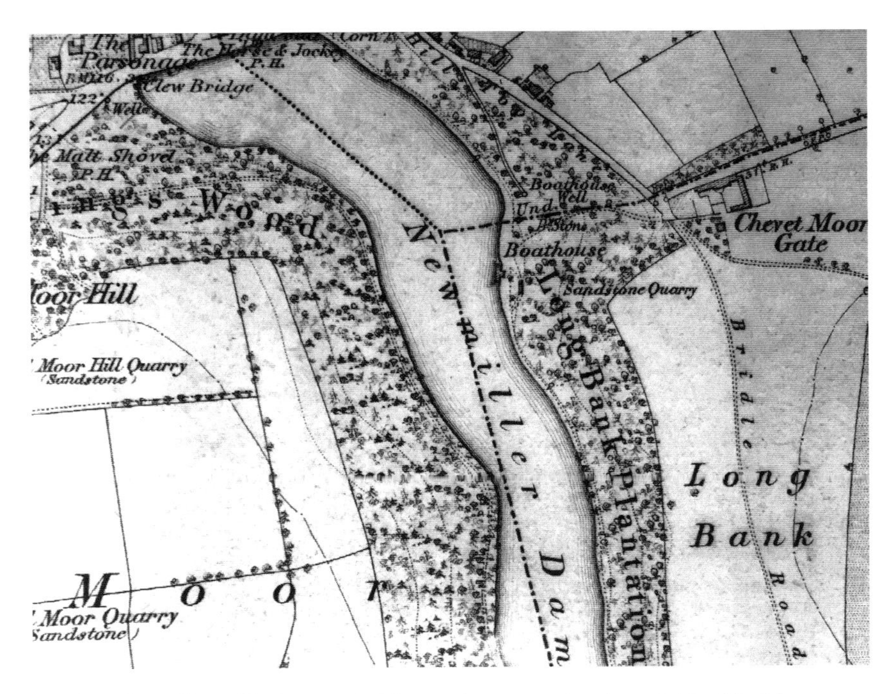

26. *Quarries at Newmillerdam, on the Moor to the west, and on the east side of the lake*

far as to have the same, for what should be rather looked upon as an Acknowledgement of the Dukes right, than upon the Footing of a Merchantable Bargain'. Marsden replied that, 'as the gentlemen of Halifax had done something formerly which had interfered with the profits of the Duke's Manour', the Duke would be less inclined to do any favours, but he would do his best. However it seems that 'his best' was not sufficient, and it was decided that the commissioners should not rent the quarry, but might buy stone directly from the quarrymen who were already working there.[92]

Despite these problems the stone was deemed the best to be used in the lock at Horbury Pasture and in April 1760 a road was being prepared for carriages to transport the stone. One of the landowners locked a gate and refused access unless he was properly compensated. The delay in sorting this out caused concern.[93] Negotiations with Mr Marsden, whom Smeaton claimed was reckoned 'a cunning man', were continuing in July 1760, when Smeaton in a 'long conversation' insisted that the commissioners would only pay the occupiers of the quarries wherever it could be cheapest procured.[94] There is no mention of stone from Newmillerdam in the Journal, but the accounts show that stone was purchased from individual quarrymen at Newmillerdam in August 1761.[95]

Meanwhile other decisions had been made:

> 'Having with Mr Nickals and Mr Wilson resurveyed Principal Quarries ... we find the best to be the New Millar Dam, Rob Hartly's stone at the out-Woodside and Widdow Hartly's stone at Altofts; from all of which places tenders have been Offered at nearly the same price, but that none of those Sort are perfectly proof against weather both in and out of Water; however as none of these are remarkably defective nor can any place furnish our whole Quantity, and being the best we are likely to get, we propose to make the best contracts for the same that we can, and endeavour to sort and Apply them to their proper Uses, so as to Answer the End as well as possible, we also Judge the blue Stone Quarry near Netherton to be the best sells, Hollow Posts and Coins and other Stone for particular Uses, where great Strength and closeness is required and shall endeavour to in due time to contract for the same.'[96]

The site of Robert Hartley's quarry at 'Outwoodside' is unclear, he is also mentioned as having a quarry at Lee Moor which is shown north of Canal Lane on the 1850s map; this is on the edge of the Outwood area. He delivered stone to the 'steanard', the area near

---

92 WYJS/CA MIC2/4 11th Jan 1760.
93 WYJS/CA MIC2/4 29th April 1760.
94 WYJS/CA MIC2/4 21st July 1760.
95 WYJS/CA MIC2/22 Aug 1761.
96 WYJS/CA MIC2/4 11th Jan 1760.

Chantry Bridge, although this archaic Yorkshire term was used for any area where debris had been cast up by the river in flood.[97]

Nickalls and Wilson visited Mrs Ann Hartley's quarry at Altofts in April of 1760 and Nickalls commented in a letter to Simpson, clerk to the commissioners, that they 'found her to have been very Diligent in preparing Ashler for the Works, having already got ready near 1000 foot which is to be delivered at Fall Ing near this Place at 6p per foot She proposes next Week to deliver near 800 foot, says that Money Runs short with Her therefore begs the Gentlemen wou'd order her Ten pounds on account As she is not dependant on the Duke And as a due encouragement of such Independants, may render us the less Subject to the Quarry Acknowledgements Ask'd by Mr Marsden'.[98] It must have been an advantage to treat directly with owners, rather than negotiate with agents.

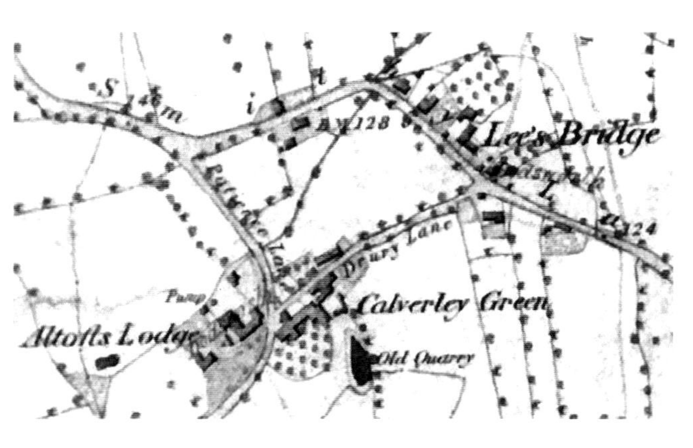

27. 'Old Quarry' at Altofts

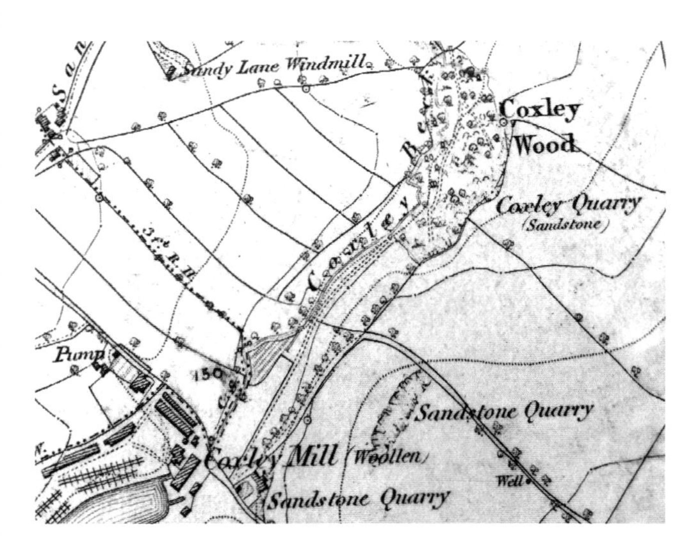

28. Quarries at Coxley and Netherton

The quarry at Netherton was in the Coxley Valley, where they were paying individual quarryman.[99] Several quarries are shown on the 1850s map, and these were close to Horbury Bridge and the Calder.

97 George Redmonds, *Names and History*, Hambledon and London, 2004.

98 WYJS/CA MIC2/4 3rd April 1760.

99 WYJS/CA MIC2/16 5/6th Nov 1760 'at *Netherton, or as Workmen chuse to call it the Coxley*'.

A letter from Nickalls to Simpson dated 17th November 1760 illuminates further how the rights to quarries were negotiated. Addingford[100] Quarry was on land belonging to the township of Horbury, and Nickalls met with the Vestry convened there. They were asking for 4d a yard for the whole surface, and that when the contractors for the Navigation had finished with it, that it should be returned to their ownership, even if some stone remained unquarried. The commissioners wanted it to become their indisputable property which could be worked at any future time, and suggested that the negotiation should be taken to a Jury (as delineated in the Act of Parliament). Eventually agreement was made with the commissioners paying 3d a yard and gaining all rights to the land quarried at that rate, and to any depth. They were also to pay any damages that were incurred in the stone being carried away.[101]

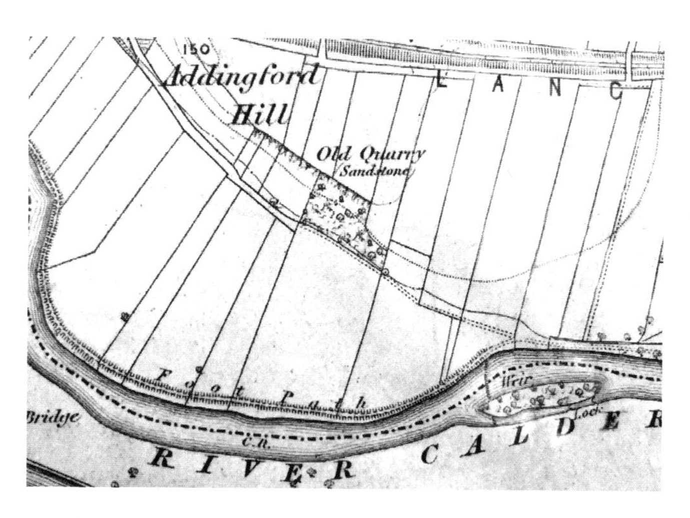

29. 'Old quarry' at Addingford

Stone from other quarries was being obtained: Agbrigg,[102] Barnsley Road in Wakefield,[103] Carr Gate[104], Lee Moor[105], and further afield from Huddlestone.[106] On the other side of the river to Addingford stone was being quarried at Hartley Bank.[107]

As work proceeded upriver stone was obtained from Mill Bank close to the river east of Thornhill.[108] By June 1761 the quarrymen had been discharged from Mill Bank, Netherton and Addingford quarries, presumably sufficient stone had been got for the purpose.[109] However in July[110] there was mention of troughs, and wedge courses being made at Addingford, and pieces for the conduit cloughs at Netherton – this would be work

---

100 Addingford is the modern name, the Journal refers to the place as 'Addingforth'.

101 WYJS/CA MIC2/16 17th Nov 1760.

102 WYJS/CA MIC2/16, 7/8/9th May 1761.

103 WYJS/CA MIC2/16, 5/6th Nov 1760.

104 WYJS/CA MIC2/16, 2nd June 1760.

105 WYJS/CA MIC2/16, 27th June 1760. This may be the quarry to the north of Canal Lane probably belonging to Robert Hartley, or other quarries to the north of it.

106 WYJS/CA MIC2/16, 28th June 1760. This was on the magnesian limestone ridge near Sherburn in Elmet and was of an excellent white quality.

107 WYJS/CA MIC2/16 1st July 1760.

108 WYJS/CA MIC2/16 7/8/9th May 1761.

109 WYJS/CA MIC2/16, 5th June 1761.

110 WYJS/CA MIC2/16 8th, 9th, 10th July 1761.

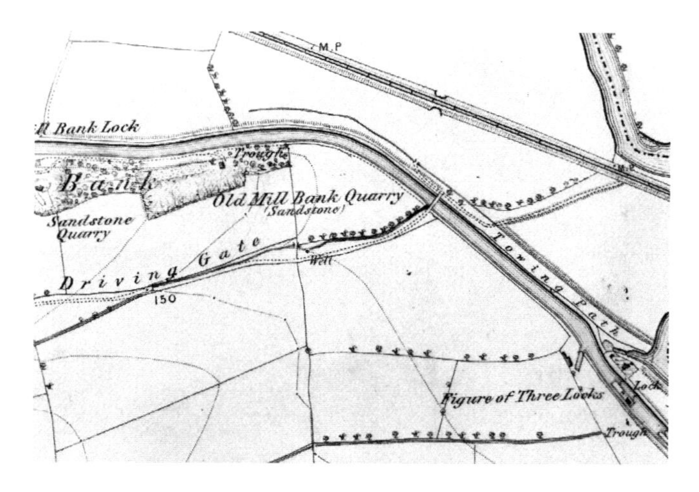

30. 'Old Mill Bank Quarry'

for the masons. Other quarries up river may have been used as they were needed, but no specific mention is found in the Journal or letters.

Apart from Robert Hartley and Ann Hartley who appear regularly in the accounts and other records, William Charnley & Co. are also named in the accounts from April 18th 1760 onward as supplying stone, although there is no reference to where they were quarrying. Later he also became a major contractor.

Problems with supply, and the employment of the masons continued throughout 1760. By July Smeaton was reporting that 'Masons were very scarce & not as ready to enter with us as might be expected.' The contractors too were 'requiring an exceeding deal of talking to & looking after, agreed also to recommend it to the committee to engage proper Quarries, & to employ therein the most usefull of our Hands during the Winter in getting Scapelling[111] & hewing Stone for the next Year's Service.'[112] In December 1760 Smeaton was still commenting that the seven stone contractors were causing 'the greatest vexation difficulties and obstruction that we have hitherto met with ... together with the Difficulties of getting or retaining any number of tolerable Hands for Summer work only without being able to employ them in winter'. Smeaton determined that they would take control of finding their own quarries and obtain the stone at a better price: 'this step is likely to turn out so well that we shall be able to furnish ourselves with sell pieces and Hollow posts not only of a much better quality but considerably cheaper then we have hitherto been able to do; and that our aisler and Damstones[113] will come Cheaper by 1d per Foot and our backing by at least 1/6 per Yard in 5/0s as appears from such tryals as have already been made: and in order to secure our selves in this matter we are now endeavouring to reduce the whole to task work as fast as we can come to the true value of each part; and this I am convinced is the only way by which the work can be done well and upon the cheapest Terms'[114].

---

111 Scapelling- worked stone.

112 WYJS/CA MIC2/16, 15th July 1760.

113 Victoria Owens (Newcomen Society) suggests that 'damstones' were the large stones used in the construction of the weirs.

114 WYJS/CA MIC2/4 6th Dec 1760.

The working arrangements therefore changed with the quarries at Barnsley Road, Addingford and Netherton worked directly by the commissioners' men with masons to do piece work, preparing stone at rates listed in the December 1760 journal. Masons were paid by the yard or by the foot, depending on the type of stone they were working, for instance at Addingford in December 1760 they were paid 2d per foot for 'aisler', 2d per foot for damstones, 1s 6d per yard for scappelled backing, and 2d per yard for wall stones.[115] The accounts show that from June 1760, payments continued to be made regularly to the quarries and quarry owners, but wages were also paid to individuals 'on the stone acct'.[116] From 1762 responsibility was handed over to the major contractors.

Joshua Wilson resigned on 27th November 1760, possibly owing to the difficulties already described in sourcing stone and finding masons, but continued to be paid until Christmas.[117] The post of Master Mason was not refilled. However in July 1761 his name reappears in the accounts and the considerable sums he received regularly suggest he had taken on a role as a contractor. In November 1762 he was contracted to build the locks at Cooper Bridge, Kirklees Mill, Anchor Pit, Cooper Bridge Dam and Anchor Dam for £1230 and to get stone near Cooper Bridge.[118]

Without a Master Mason throughout 1761, it seems that no one person was made responsible for oversight of stone supply and working, perhaps adding to Nickalls' workload. Following the dismissal of Nickalls in November 1761, at a meeting of the commissioners on February 17th 1762, it was ordered that an advert be inserted in the General Evening Post in the York, Sheffield and Manchester Papers: 'Wanted two Surveyors for the Calder Navigation one to superintend the Masonry and Diggers, the other the Carpentry; both under the Direction of Mr Smeaton. On 31st March 1762 it was resolved that 'Mr Mathias Scott be elected surveyor and superintendant over the Masonry and Digging with a Salary of £60 per Annum to commence 1st April 1762'.

Visits to quarries, bargaining with owners of quarries, and quarrymen themselves, the scarcity of masons, and the vagaries of the weather which prevented the transport of the stone occupied much of Smeaton's and his subordinates' time. The pressure to succeed in as short a time as possible, and as cheaply as possible must have been immense.

Wood supplies

The supply of wood presented the same transportation problems as stone. Imported wood, usually firs from Northern Europe, could use the Aire and Calder Navigation to Wakefield but any local source had to be close to the town or near water transport as the

---

115 WYJS/CA MIC2/16 12/13th Dec 1760.

116 WYJS/CA MIC2/22.

117 WYJS/CA MIC2/2 27th Nov 1760.

118 WYJS/CA MIC2/2 18th Nov 1762.

large quantities required would be both bulky and heavy to carry. The quality and sizes, particularly of oak, needed to be carefully checked before pricing and ordering, involving lengthy visits to view and select timber. Trees were a growing and changing material and even after cutting down there were further processes of removing bark (pilling or peeling), and seasoning before or after sale. The navigation was always in competition for the large oaks with the King's ship builders and other boat-building. All these considerations are clear from the lengthy and sometimes problematic supply arrangements made for the navigation.

As soon as the Act was passed the commissioners began to focus on the sourcing of materials. In a letter of 10th July 1758 they were asking Smeaton what he needed for the work, not only the mechanical tools and instruments, but also all materials. In reply from Plymouth he described the types of wood required: 'a Quantity of the best Oak timber, the sooner the better as there will be more time to season it before wanted. Elm and alder will also be useful if to be purchased under the price of Fir: as also Oak poles from 5 to 9 inches Diameter for Piles; which ought to be well seasoned in the Water before used. All kinds of Fir timber I apprehend can be had as it is wanted.'[119]

Smeaton was empowered by the commissioners in November 1758 to begin to source the materials he required[120] and immediately two Tadcaster merchants, Francis Iles and Robert Fretwell offered timber. Iles, who appeared to have some understanding of their requirements, wrote to the clerk, Bentley, on 20th November. 'I ... can serve them with a many large Oak Trees proper for the Use ...from 80 feet in measure to 120 or 150 feet. They are from two large falls of Timber I have going down at Ackworth dark and Nostel, a part of which is now lead to the water side at Knottingley.' Rough prices were explored by Bentley on behalf of the commissioners. However Iles pointed out that: 'prices vary according to lengths, thickness and shape of trees so can't give any standing price until the proper assortment is fixed upon. I have £5000 or £6000 worth on hand and in it timber of any price ... A medium ... for an assortment you require may be about 18d to 2sh a foot, the Large is now wanted for King's service. I can deliver at Wakefield but as yet have not considered the charge of delivering there.' His letter indicates the many considerations in buying wood.

In late 1758 Smeaton arranged with John Horn, the millwright from Wakefield, to view Robert Fretwell's timber, and Smeaton himself went to view it when he was returning from Scotland in December. 'I have no reason to be unsatisfied with the wood either as to quantity or quality yet the trees are in general too short lengths to answer the principal purposes for which they are wanted.'[121] Anything useful would therefore need sorting out. His recommendation was that there should be no general contract with Mr Fretwell at that

119 WYJS/CA MIC 2/4 15th Aug 1758.
120 WYJS/CA MIC 2/2 15th Nov1758.
121 WYJS/CA MIC 2/4 15th Jan 1759.

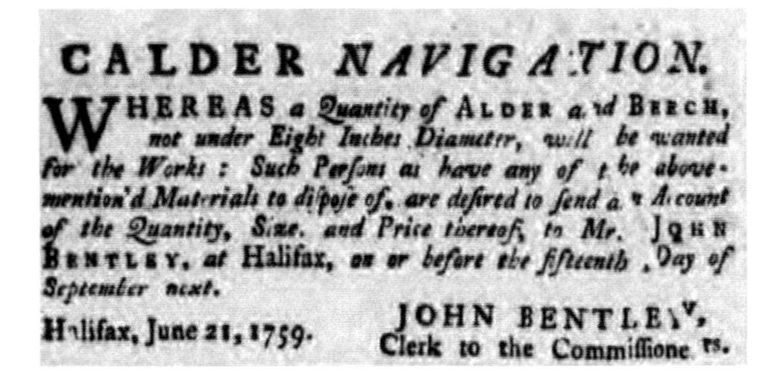

CALDER *NAVIGATION.*

WHEREAS a *Quantity of* ALDER and BEECH, *not under Eight Inches Diameter, will be wanted for the Works: Such Persons as have any of the above-mention'd Materials to dispose of, are desired to send an Account of the Quantity, Size, and Price thereof, to Mr.* JOHN BENTLEY, *at* Halifax, *on or before the fifteenth Day of September next.*

Halifax, June 21, 1759.

JOHN BENTLEY,
Clerk to the Commissioners.

31. Leeds Intelligencer 10th July 1759

time and he had written to Fretwell to let him know. The commissioners agreed with this view.[122]

In the middle of the following year the commissioners advertised[123] and more suppliers of wood responded. On 5th July a Mr Wormald offered to supply 'oak, elm and ash, shippable', but he made clear the navigation was in competition with ship builders and he was due to go to the seaport towns for his orders. William Foulis near Northallerton offered alder on July 13th but requested that his wood should be viewed to assess its suitability. He would be at the expense of cutting down and transporting to the waterside at Stockton-on-Tees, a distance of twelve miles. All offers, Bentley replied, would be considered on October 3rd at a commissioners' meeting. In reality they still had no-one to inspect and make decisions about particular requirements as they had not employed Nickalls at the time Smeaton suggested. Neither did they have any land to house the materials until the following year.

It was not until 4th December 1759, following the commissioners' meeting of 28th November at which Smeaton was in attendance, that sourcing wood began in earnest. A letter from Thomas Simpson, the new clerk to the commissioners, was sent to all the suppliers who had been in touch: 'The Committee of the Calder Navigation desire you wou'd inform them how far you are able to serve them with Oak not under '12 Inches Dia' Alder and Beach not under 8 Inches, and where they lye, and at what per foot you can lay them down at Wakefield.' The potential suppliers at this stage were Mr Francis Iles, Tadcaster; Mr Fretwell, Tadcaster; and Mr Wormald of York; Mr Richard Milnes of Flockton, and Mr Joseph Lee of Mirfield. Although the two latter names had not featured in the previous exchanges of letters, they were local men and would certainly have been known to some of the Halifax gentlemen.

Mr Lee of Shillbank between Dewsbury and Mirfield, replied first on December 6th, offering oak and beech, both peel'd and unpeel'd, but he felt the transport to Wakefield would be difficult 'laying them down at Wakefield - I cannot tell how that may be for they will be a great Way to hurry and teams are dear to us'. Mr Fretwell replied from Potterton near Barwick-in-Elmet on 8th December. He mentioned in an aggrieved tone that 'I did this Time 12 Month press much for to sell then about 300 Tuns of Timber which they gave

122 WYJS/CA MIC 2/4 17th Jan 1759.

123 LI 21st June 1759. These advertisements remained in the paper for three or four weeks.

me great Encouragement and on that account put myself to some Expence and Trouble in going ... with Mr Smeaton and his Agent to view my Timber, and a Certificate were given me both of the Quantity and Quality ... being proper for the use of the navigation.' He also thanked the committee 'for their giving me the preference' although there appeared to have been no decisions made at this point. On 10th December Joseph Wormald of York believed he had '400 Tons of fine Oak wood yet left of dfft Lengths and Quality, which must consequently be of different Prices.' He too asked for someone to look at these and give him an idea of what sizes and quality they needed. The last reply came from Mr Iles on December 18th. He sent another detailed letter with an account of his oak 'of which large trees and fit for your purpose either for Plank or any other Uses about the Locks', offering to contract for the timber whole or to be converted in his yards. He wanted to meet Smeaton at Ferrybridge to visit his yards but he expected 'to sell my large to the King's Use for shipping, although they take nothing under 40 feet. The Oak above is now ready for Use and at the Riversides and I shall have more to Fall the next Pilling Season.' He also offered 'Petersberg Deal' and 'Gothenborg iron'.

On January 2nd the commissioners minuted that Mr Smeaton and Mr Nickalls would be responsible for the timber supplies as well as other materials. Early in January the pair travelled to York calling to see Mr Iles at Tadcaster and Mr Wormald at York but no decisions were taken then as Nickalls had not seen all the wood stocks.[124] On his return Nickalls went to Gainsborough to see Mr Thomas Holmes who thought he had a large quantity of wood near Knottingley which would therefore be easy to carry to Wakefield. This assessment of the timber offered in the tenders took Nickalls considerable time during this extremely busy period: on February 25th he charged for four days' expenses 'after timber at Mr fretwell's 10sh'.[125] On this trip he may also have checked Iles' timber. There is, however, no evidence of the reasons for his final choices of suppliers.

By 1760 the commissioners were looking for a foreman of the carpenters and Nickalls recommended John Gwyn, another Londoner. On 26th January the commissioners received Gwyn's confident reply to their employment offer in which he negotiated his salary. They suggested 18 shillings for each week worked, but he expected a regular salary for every week of the year at that rate as his 'fix'd' requirement. He was to be paid from the day he set off from London, and all travelling expenses then and on any other Navigation business were to be met. The commissioners accepted his terms and requested that in March he be on site at Wakefield where a workhouse and storehouse with another shed for the sawyers were being prepared in the Wakefield yard.[126]

Wood purchases can be seen in the accounts compiled by Nickalls, which were drawn up to send to Mr Simpson from January 1760. The first purchase of wood was for 270

124 WYJS/CA MIC 2/4 23rd Jan 1760.
125 WYJS/CA MIC 2/22 25th Feb 1760.
126 WYJS/CA MIC 2/16 26th May 1760.

stakes, made by Luke Holt, a local carpenter, for staking out the cuts and locks, a task which Smeaton and Nickalls immediately completed. Nickalls' purchases from Hull in January followed: considerable quantities of fir and deal timber from Ben Blaydes and Messrs Hugh Lawson and Co. Such purchases were made by letters of credit which proved a problem when Nickalls, an unknown person to the northern merchants in Hull, was not provided with a letter of credit. Eventually he applied to Sir George Savile, who must have been in Hull at the time and who organised the transaction on his own credit. In a letter to Simpson addressing Nickall's difficulty, Smeaton lectured on prompt payment:

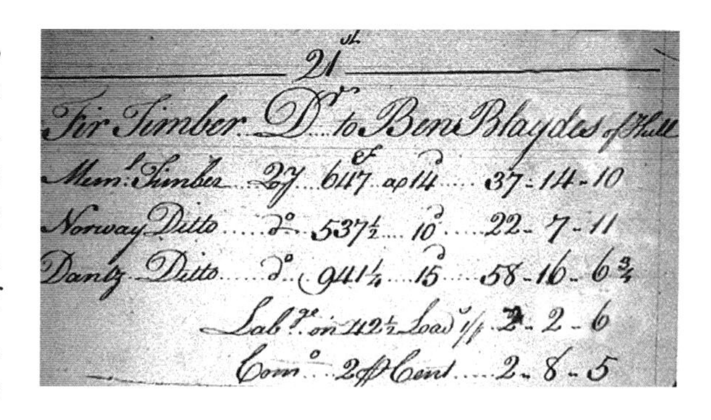
21
Fir Timber Dr. to Benn Blaydes of Hull
Mem: Timber ... 27 ... 647 ... a 14 ... 37-14-10
Norway Ditto ... Do. ... 537½ ... 10 ... 22-7-11
Dantz Ditto ... Do. ... 941¼ ... 15 ... 58-16-6¾
Lab. on 42½ Load 1/- ... 2-2-6
Com. 2 p Cent ... 2-8-5

32. Payment to Benjamin Blaydes for timber, 21st January 1760

> 'I am convinced that this must happen either by some Mistake or unintended Omission yet I cannot help observing that those are particularly unlucky, when they happen at the Commencement of a public Undertaking: the Credit of which is of a more ticklish Nature than even that of a Private Man, because public Bodies are of a Less determinate nature, and if any thing happens amiss, not so easy to be dealt with or a Satisfaction to be obtain'd, Nothing but Ready Money and punctual Dealings can raise Credit of the Sort.'[127]

On April 3rd Nickalls sent news to the committee that he had bought £153: 18s: 6d worth of timber, mainly elm, from Mr James Milnes of Flockton. In the same letter he announced that: 'I went from Halifax to Mr Rob Fretwell's ... and did agree with him for one hundred and fifty tons of oak to be delivered at 19d per foot at the store yard which will come to £447, which sum is to become due at one Month after delivery and that he proposes to deliver the same in a Fortnight.' From late March Fretwell's oak timber was arriving regularly in the yard and in the Journal's first entry in May Smeaton observed that there was 'a considerable Quantity of good Oak, Elm, & Fir Timber with some Alder for Piles'. On July 17th Nickalls compiled his account of the whole order from Fretwell, showing that deliveries and payments were completed apart from £44 worth of oak. Other timber arrived from different sources: on July 2nd 1760 Richard Milnes, son of James from Flockton, supplied £140 worth of oak, alder and elm, and Thomas Cotton at Haigh was paid £139 for elm on August 30th. By September Nickalls warned that fir timber was lacking and 'must be had soon or the works stand for want thereof'. It arrived in Hull on 21st October and different sizes and dimensions of 'common deals tough, redwood tough,

---

127 WYJS/CA MIC 2/4 2nd Feb 1760.

small baulks tough' were bought from William Joliffe, timber merchant.

However by the end of the summer of 1760 a problem with Robert Fretwell had arisen. Nickall's initial order for oak from Fretwell seemed to have been received and paid for before July 1760, but he was asked to prepare his accounts again because of Fretwell's complaints. In October Nickalls was asked to send Simpson the letters concerning this order and the receipts of timber into the yard. Meanwhile Fretwell refused to meet with Nickalls to sort out the problem. Fretwell's argument seemed to be that Nickalls had ordered or promised to order much more wood than the 150 tons delivered that year, and on 19th January 1761 he sent more timber by boat to Wakefield which Nickalls refused to accept.

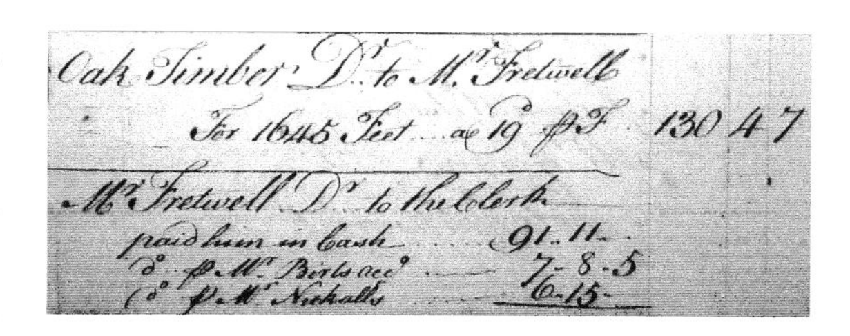

Oak Timber Dr. to Mr. Fretwell
For 1645 Feet ... at 19 p F. 130 4 7
Mr. Fretwell Dr. to the Clerk
paid him in Cash ... 91.11.
Do. p Mr. Birls acct ... 7.8.5
Do. p Mr. Nickalls ... 6.15.

33. Payment to Fretwell for timber, 9th August 1760

Fretwell intensified his onslaught. At the commissioners' meeting of 27th January 1761 it was announced that 'Mr Robert Fretwell hath commenced an Action against some of the Commissioners' and on March 3rd they agreed to refer the dispute to 'two gentlemen of character'. Simpson hoped this would settle the problem 'amicably without the great expense of Law'. Arbitration Bonds were entered into and the meeting was arranged for Wednesday 25th March to meet at the White Hart, Wakefield.

At their meeting of 2nd April 1761 the commissioners heard the outcome of the White Hart arbitration. Samuel Lister of Horton and John Hill Esq. of Oxton had heard the evidence on each side and had nominated Thomas Cotton Esq. of Haigh the Umpire. They listed what had been agreed as a result: Fretwell could deliver 200 tons of oak timber which was now lying at Allerton Bywater but only if the timber was of certain specified dimensions to be checked on arrival by Gwyn, as Fretwell refused to deal with Nickalls. Fretwell was also given £20 towards damages sustained when he had tried to deliver the wood in January and another £20 towards his costs. It appeared that to some extent he had got his way. 'If I was unacquainted with Fretwell's character I shou'd be amaz'd at his Impudence.' Smeaton wrote on March 3rd, suggesting that he had not been properly confronted.

After the Award it was arranged that Fretwell could begin to deliver the wood, although Simpson made it clear that 'the Timber as soon as d[elivere]d shall be measured ...' and the commissioners would pay for 'such Quantity as shall be jointly certified for by Mr Gwyn and your Agent provided you have no Objections to referring to Mr Smeaton (who will be in Yorkshire in a fortnight or 3 Weeks) whether the Timber so deliver'd be such as is directed by the Award. They hope you cannot reasonably object to this Method'[128]

128 WYJS/CA MIC 2/4 3rd April 1761.

However from the first boatload Mr Gwyn refused to take some of the wood and Fretwell immediately complained: 'So long as Mr Gwyn is under Mr Nickalls it will be hard for him to do me Justice', Simpson's reply was that Gwyn's decisions needed to be checked by Smeaton. In response Fretwell left this wood and more outside the yard but by the river in Wakefield. He stated that he would 'look upon it as actually deliver'd' thus getting 'quit of it in such a manner as will be binding upon the gentlemen to accept my bills of value.' His invective against Nickalls became ever more unpleasant: 'A £100 would not pay the injury I have suffer'd, ... ; my words and Applications was despised when a stranger was believed ... if I get no answer I shall load 2 Vessels tomorrow and take care so as nothing on my side be undone according to the award. If Mr Nickalls be by the gentlemen appointed to take it in I shall regard no certificat'. When he arrived with these boats he wrote complaining that he had been ignored by Gwyn at the behest of Nickalls. 'Such an Instance never was of such Treatment'.[129] He left this cargo on the Stennard below Wakefield Bridge.[130]

As soon as Smeaton arrived in May 1761 he inspected the rejected timber and sent a letter to Simpson, countersigned by Nickalls and Gwyn. He wrote with great clarity and authority about the poor quality and apparent recent felling of this wood. Fretwell's rather peevish reply was: 'I will send Old Timber such as shall fulfil the Award & if not so proper for the Works in shape as that Timber is now laid down in Wakefield I am with safety vindicated my so doing and all from the Stupidity of Managers'.[131] The account book records the deliveries of oak into the yard from Fretwell on 2nd April of £169 paid on 22nd April, and the second on 27th June of £253. There was a small bill in August 1761 after which the accounts show no further payments to him.

Other sources[132] show that Fretwell was in financial difficulties at this time which might explain his increasingly frantic and aggressive efforts to dispose of his wood. He had been a successful tanner who appears to have overstretched himself in property and land transactions, and in his transport businesses. By the early 1750s he had various lighters and sloops using the River Wharfe at Tadcaster which was navigable to the Humber, and a carrying business with wagons and carts between Hull and Leeds.[133] However in 1757 his creditors had appointed trustees of his estates and effects, although they had left Fretwell himself in charge of the transport business 'subject to the Trusts'. In the same year much of his land and property was sold.[134] By July 1762 his trustees were advertising to let the Navigation and the mills, grounds, wharfs and warehouses at Tadcaster, and at the end

129 WYJS/CA MIC 2/4 10th/14th April 1761.

130 WYJS/CA MIC 2/4 21st May 1761.

131 WYJS/CA MIC 2/4 25th May 1761.

132 www.barwickinelmethistoricalsociety.com.

133 Manchester Mercury 23rd June 1753, Fretwell's advertisement for his carrying businesses.

134 LI 30th August 1757.

of that year Fretwell sailed for India to restore his fortunes, dying there in 1769. A later obituary stated that he had been 'Esquire of Potterton, Near Tadcaster, in Yorkshire, well known in that county for his extensive connections in trade, in which, meeting with a continued series of misfortunes, he determined to quit his native country ... with a fixed resolution to re-establish his family in that affluence from which his many heavy losses in trade had reduced them.'

Wood supplies continued. In the summer of 1761 Nickalls agreed supplies of oak, alder and ash timber worth £39 from a Wakefield gentleman, William Richardson, explaining his reasons: 'its laying so convenient for Lupsit Dam, and that tho' the Oak is not so prime clear Timber as some, yet its as hearty as any, and for the Bottom Work of Dams etc it is as good as that of double the Price'.[135] There were also further timber supplies from Richard Milnes for which he was paid £22 on 27th June and £112 on 4th July 1761. Another timber supplier, John Saunders, was paid £93 on 16th May 1761. Later timber suppliers become more difficult to pick out in the accounts.

It seems very unfortunate for Nickalls that Fretwell has caused such trouble. In whatever way the affair was resolved it left a question over Nickalls' management. Additionally Fretwell had branded Nickalls a 'stranger', brought into the north by Smeaton, a view which may have resonated with some of the commissioners and reminded them of their original opposition to Nickalls' role. However the Fretwell affair had thrown up another question mark. In a letter of 30th May to Simpson, Thomas Cotton, the umpire in the Fretwell case, revealed that during the adjudication it was mentioned 'that Mr Nickals had insisted upon a Farthing a Foot being allow'd to him by one Mrs England, as a Perquisite of some Wood he was going to pay for'. This question of Nickalls' integrity became widely known and prompted testimonials from other suppliers: 'As I am not Ignorant what some People say of Mr Josh Nickals'[136] wrote Mr J Wilkinson of High Burton 'I take this Oppy to acquaint you that he was punctual & honest', and Richard Milnes of Flockton assured the commissioners that 'Mr Nickals in all his Contracts with me which were not inconsiderable never insinuated any such Thing to me'. However this further suspicion could have brought about the sudden dismissal of Nickalls which was decided at a commissioners meeting on 18th November 1761, when they stated: 'It appears to the Comms that Mr Nickalls conduct as Deputy Surveyor is so far culpable as to deserve to be discharged.'

However, other factors also may have fed into the decision. When Smeaton was away Nickalls' workload seems to have been extremely heavy. He oversaw all the work on the Navigation, co-ordinated the buying of materials and land as the work progressed

135 WYJS/CA MIC 2/4 undated letter sent around 13th or 14th April 1761.
136 WYJS/CA MIC 2/4 1st May 1761.

upstream, employed the labour force and paid the men, kept accounts, wrote letters to Simpson about all the issues which arose in these areas, as well as to Smeaton to keep him abreast of progress. He managed all the alarms and difficulties encountered along the river, often without Smeaton's immediate input. It seems no wonder that he fell behind with his accounts: his letters often begin with an apology that he has not completed them.

On 18th August 1760 Smeaton tried to explain this: 'it has been occasioned by the constant attendance he has been obliged to give at the Lock since the foundation has been begun, but when this is over the Men will be somewhat initiated against another, he does not expect to be backward any more nor can he receive assistance therein as no Person less acquainted with the Particulars of the Works than himself can digest them into the form that they have appeared'. Nevertheless the difficulties continued especially in the winter when Nickalls was ill for periods. Eventually Nickalls wrote to Simpson: 'I find 'twill be impossible for me to do any part of the Accounts and pay a due Attention to the Works, … it possibly may be ask'd why I undertook it, I answer from a Motive to do my utmost to serve the Works'.[137] The work does not seem to have been reallocated and relations between Nickalls and the commissioners continued to deteriorate. On May 9th 1761, Nickalls found that Mr Simpson had not brought to Wakefield the money for the wages and purchases. He begged the commissioners 'that if in anywise I have made a false step that might not involve others. And as the Quarries can't be discharg'd without Cash, and as the other Parts of the Work going without may bring a Dispute on the Gentn Acting therefore most earnestly beg you'd be consenting (pray don't consider it that I mean to direct) to advance this weeks Supply'.

Following Nickalls' dismissal the commissioners began to issue direct working instructions, including those relating to a new use for wood: 'in the execution of the Rest of the Locks upon the Calder where there are sand gravel or soft foundations the present plan of constructing the Stern Sill and Recess of Stone be changed, and that in future they be constructed of Wood'. At the same time they took the decision that carpentry work was to be put out to contractors who would supply their own timber thus removing the work of sourcing and paying for the wood from the engineers. The remaining timber belonging to the commissioners was to be taken by the contractors at a

CALDER NAVIGATION.

THIS is to give Notice, that the Carpenters Work of the Locks and other Works of the *Calder Navigation*, will be lett by Meafure or by the Great, as alfo the Pile driving and Drainage of the Water. A Plan of the Lock for next Year, may be feen with Mr. SIMPSON, Clerk to the Navigation: Propofals will be receiv'd by the Commiffioners at Halifax the 16th Inftant.

☞ The Timber belonging to the Commiffioners in their Work Yard at Wakefield, to be taken by the Contractor at a fixed Price. The Nature of the Piling Work will be explain'd by Mr. GWYN, at the faid Work-Yard.

Halifax, Dec. 2d, 1761.

34. Leeds Intelligencer, 8th December 1761

137 WYJS/CA MIC 2/4 21st April 1761.

fixed price.[138] Even so, when a timber-laden flatt was carried away by floods in 1763 the timber lost was 'chiefly belonging to the Contractors but some to the Commissioners.' Gwyn became surveyor for both carpentry and smiths' work.

Regular weekly payments in the accounts went to lists of individual contractors. For example the storeyard wages including those for the carpenters and sawyers were entered in the accounts until around April 1762. After this Luke Holt, the local tradesman who had provided the first carpentry work in 1760, is recorded as Luke Holt and Topham, later Luke Holt and Co. This business received substantial payments of £10 to £40 a week for the remaining years of the Journal and may have done much of the carpentry work. The separate weekly storeyard wage bill for wood workers disappeared from the accounts.[139]

Gathering Materials and Equipment

A wide variety of materials needed procuring apart from the stone and timber already described. These are recorded both in the Journal itself, the minutes of the commissioners' meetings, the letters and reports, and the accounts. It can be seen from the first entry in the Journal on 26th May that much was in readiness to proceed; materials and equipment had been gathered, and labour employed.

Apart from stone, bricks were also bought, £3: 1s: 9d being paid in July 1760. Bricks would have various uses: constructing access chambers, flumes, walls, arches etc. Whilst stone was the predominant material it has its limitations and, with a smaller standard unit size, bricks can be more versatile for construction. Bricks were a local product and would have been readily available, for instance from Eastmoor or the brickworks on the road from Wakefield to Ossett.

From early January 1760 Nickalls was sourcing iron, anvils and and smiths' tools but was hoping for advice on bellows. On 23rd January 1760 he wrote to Simpson:

> 'Now as we shall want a Smith soon therefore would be glad they would fix on one, and take his Opinion as to the Articles just mention'd. If my Opinion concerning the person who offered when I was at Halifax might be of use, As that from what I saw of his work, and by his discourse, I should judge him a proper person'

By 28th February the smith had started work in Halifax, although his salary of 15s a week was to start on 26th March 1760. Not until 3rd April 1760 did 'Mr John Cousin, our smith and man' arrive at Wakefield and set up their work in the shop prepared for them. Cousin continued to work on the navigation being given a contract for iron work in January 1762. The smithy required coal and charcoal, anvils, hammers, bellows and 'work shades'.

---

138 WYJS/CA MIC 2/2 2nd Dec 1761.

139 WYJS/CA MIC 2/22 1762-3.

In 1760 it was noted in the letters that £87: 11s had been paid for iron from Messrs Thornton & Co, and that both cast iron and rolled plate iron would be needed. In October 1761 Mr Cotton of Haigh was offering iron 'as he has an Iron Forge at a Place call'd Kiln House situate upon the Dun to which Place he sends the Iron made at his Furnice at Haigh'.[140] Various payments for iron occur throughout the accounts, such as 'Russia Iron 60 Barrs and Sweeds Iron 45 Barrs' in 1760. A small amount of steel was also bought.

One of the by-products of an iron furnace was minion, the siftings of ironstone, which were used to make a hard cement which set in water and was used by Smeaton for the Eddystone lighthouse.[141] A large amount of this was needed and in a letter on 18th August 1760 to Simpson, the clerk to the committee, Nickalls writes 'You'll be pleased to Acqt the Gentn of the Committee that there is due to Cotton Esq at Haigh (the Gentleman who was so kind as to give the Minion, so much Valued by Mr Smeaton for the Morter way) ...'[142] Payments occur regularly for minion in the accounts; for instance in June 1760, 499 bushels[143] were bought for £4: 3s: 2d.

Other ingredients for the mortar were needed; lime was brought from Barrow on Soar in Leicestershire.[144] The details of the negotiations are typical of the careful attention paid by Smeaton to quality, price and carriage:

'Above a year agone I recommended the use of Barrow Lime for particular parts of our Locks, since I left Yorkshire I have procured some of it and have tried some experiments upon it and find it almost equal quality with the Watchet Lime[145] used at Edystone, I have changed some letters with one Stephen Squires Lime burner at Barrow in Leicestershire who informs me that place is 14 Miles from Willen ferry which is the shortest carriage to the Trent, that the price of the Lime is 2s 6d per Quarter at the Kiln

35. Excavated lime kiln at Barrow on Soar

140 WYJS/CA MIC2/4 7th Oct 1761.

141 Peter Nicholson, *The Builder and Workman's New Director*, Edinburgh 1845.

142 WYJS/CA MIC2/4 18th Aug 1760.

143 'bushels' - a measure of capacity equal to about 36 litres.

144 Information about archaeological excavations at lime kilns at Barrow on Soar can be found online: http://www.busca.org.uk/heritage/articles/village-history/local-history/lime-kiln-workings.html

consisting of 8 Winchester upheaped Bushels and that he will undertake to deliver it at Wakefield at the price of 8s 6d per Quarter clear of all charges, provided we take a Cargo for a Vessel at a Time and it be sent in Bulk, but will stow it and find Sugar hogsheads at 3s a piece, which will preserve it quick a much greater length of Time and prevent worst.'[146]

Pozzolana for the mortar was imported from Italy; this is a fine volcanic ash added to the lime mortar to strengthen its durability. A letter to Smeaton dated 20th June 1760 shows that it was brought from 'Leghorn' (Livorno) by the merchants Thomas and Bouchier Walton in London, up the coast and via the Aire and Calder. They suggested that it should be put into casks as ballast to avoid 'metage' – a fee paid for measuring amounts.[147] It had arrived by the 15th August. The accounts also refer to 'Terras' being bought from Benjamin Blaydes, a merchant in Hull,[148] and it is likely that this is 'trass' - like pozzolana, formed from volcanic dust into a rock and found in the Eifel region on the border of Germany and Belgium.

In June 1761 £2: 10s: 8d was paid for 144 bags of moss, but its exact use is unclear. It may have been used by itself or mixed with pitch to waterproof wooden pipes or the lock gates, or it may have been used to waterproof the walls of the locks as it was on the locks at Loch Ness.[149] Alternatively it may have been used as a roofing material.

Machines and equipment were bought from the Eddystone. On the 15th August 1758 Smeaton wrote to the commissioners, 'Respecting tools, Instruments etc we use various kinds of tackle in the Eddistone works for lifting and moving large stones etc which after the work is compleated will be sold to the best bidder, I apprehend these would be very serviceable in the works of the Calder, And tho' some of them may then be far worse, yet as the price will be proportionable, they will be worth having to serve as models to make others by. What will be chiefly wanted further at the beginning will be 2 or 3 engines of different sizes for driving Piles and Pumps of different kinds and sizes for draining of water.' By November the commissioners had agreed that Smeaton should pursue this.[150]

William Jessop was also purchasing equipment in Plymouth for Smeaton, and he lists triangles, shears, chains and other items in a letter to Mrs Smeaton on 13th August 1760. Apart from the equipment from the Eddystone, much was made on site in the workshop

145 Watchet (Somerset) lime contained magnesium and iron compounds making particularly good hydraulic mortar. (Ref: Watchet Conservation Society newsletter 43 Nov Dec 2015)

146 WYJS/CA MIC2/4 22nd April 1760.

147 WYJS/CA MIC2/4 20th June 1760.

148 WYJS/CA MIC2/22 March 1760.

149 Suggestions by members of the British Bryological Society (dedicated to the study of mosses and liverworts)

150 WYJS/CA MIC2/2 15th Nov 1758: 'Resolved that Mr Smeaton is desired and he is hereby Impowered to pursue such of the Tackle at the Eddystone Lighthouse as will be necessary for the purposes of the Navigation.'

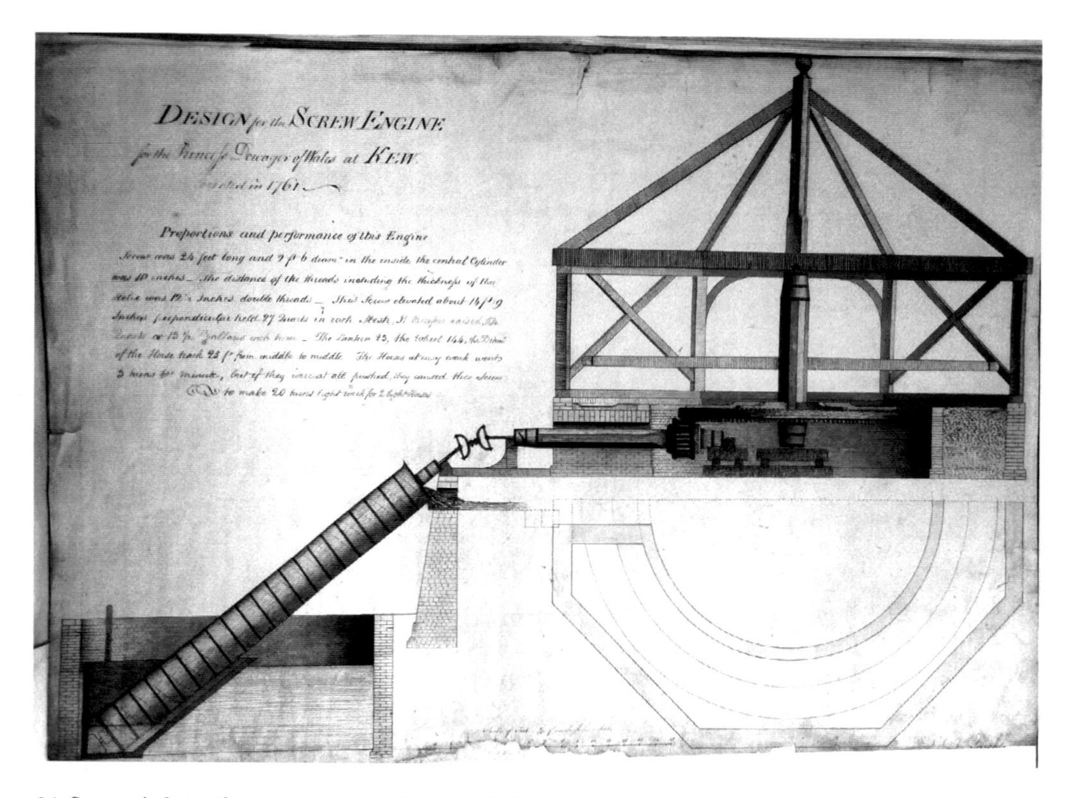

36. Smeaton's design for a screw engine at Kew, for the Princess Dowager of Wales 1761

and smithy at Wakefield to Smeaton's designs: screw engines, hand pumps, triangles, pile engines, frames for horse gins, punts to carry the engines and other equipment, a mill for grinding minion and pozzolana and a 'moveable Cabbin to hold Tools Instruments Papers while the Locks are building.'[151]

Twelve grindstones were paid for in March 1760[152] and high quality 'French stones' were bought for the mill.[153] Wheelbarrows and carts were needed. Oil, tallow, tar, brushes and brooms are also mentioned and 'hair' - although what type of hair it was and how it was used is not clear. Recorded too are payments for ropes, nails, measuring chains, rosin and beeswax, amongst other commodities. Horses were bought for the horse gins, and fodder to keep them fed. Some items are not detailed but recorded under the general heading of 'tools' or 'utensils'.

The workshop at Wakefield was busy through the winter of 1760-1761 and on Smeaton's first visit at the beginning of 1761 he listed the equipment that had been made:

'At and from the Work Yard the following Machines etc have been produced. Viz.

---

151 WYJS/CA MIC2/16 10th June 1760

152 WYJS/CA MIC2/22 March 1760.

153 WYJS/CA MIC2/16 3rd, 4th, 5th June 1762. French burrstones, formed from freshwater quartz quarried at La Ferte sous Jouarre in the Marne Valley, became world famous.

Pestle and Mortar Mill

Smeaton's design for a screw engine for raising water, grinding materials and beating cement by horses. 1760

Inscribed in the hand of John Farey: 'This is mentioned in a letter from Mr Smeaton to Mr Nichols dated 24 April 1760'

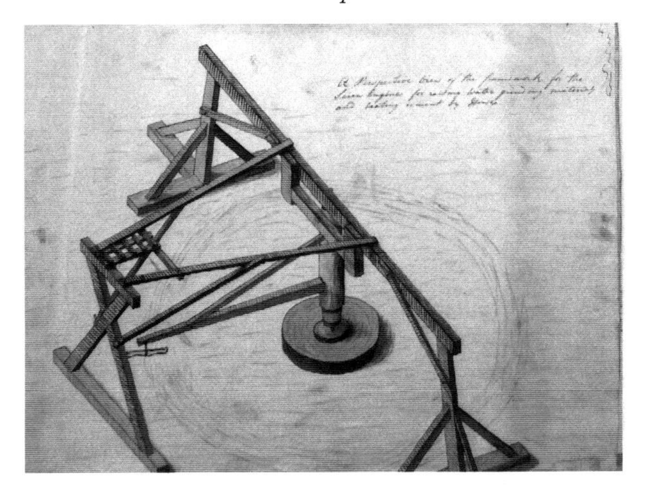

37. Perspective View of the framing

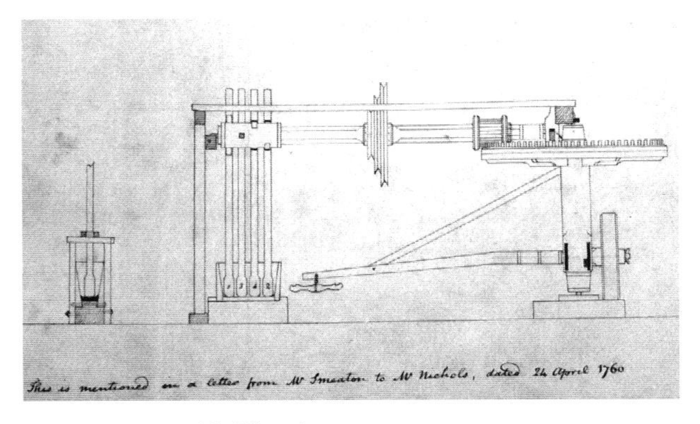

38. Elevation

1st A Floating Stage with Screw Legs, for supporting the Engine, in driving gauge Piles for the Dams; employ'd at present at Lupset
2nd Three new Piling Engines
3rd Two small Water Screws to go by hand, with Frames etc to D[itt]o.
4th One large D[itt]o and frame, with horse Frame & Stampers[154], not quite compleated, but going on with all Expedition
5th A Horse Frame for the 2 feet Water Screw ready for Setting up
6th A laving Wheel to go by hand not quite compleated.'

Abraham Rees's Cyclopedia[155] describes a 'laving gun' as a trough five or six feet in length, with a small end and a large end, pivoted on a bank to scoop the water up and feed it over the bank through the narrow spout. Several of these could be fitted together to make a wheel.

Payments for the transport of equipment, stone, timber and other materials by water and by land were paid. It was estimated on 31st December 1760 that the cost of freight would be £65: 16s 1d.

The administrative side of such a project required stationery, postage, wages books, payments for advertising, and printing. In June 1759 the committee had ordered that a chest should be procured, 'for the Purpose of keeping the Books, Papers and Records of the

154 Victoria Owens (Newcomen Society) suggests the 'stampers' are the pestle part of the stamping mill.

155 Abraham Rees, *The Cyclopedia; or Universal Dictionary of Arts, Sciences and Literature* Vol 38 1819.

said Navigation with a strong Lock and two Keys',[156] and in October 1761 a payment was made for 'drawers for the office'.[157]

Apart from the materials and equipment the accounts show payments to contractors, to quarries, to the smiths, and wages to individuals. Drink money for the labourers was an expected expense. When Nickalls was appointed as Superintendent, an allowance of £21 was made for him and his family to move to Yorkshire.[158] Other expenses that were accrued in travel and board and lodging were also remunerated.[159] As the buying of land proceeded the commissioners were also liable to pay land tax, and parish levies.

An estimate of costs was drawn up by the committee at the end of the first year and sent to Smeaton on 28th February 1761.[160] It provides a clear idea of the outgoings in that first

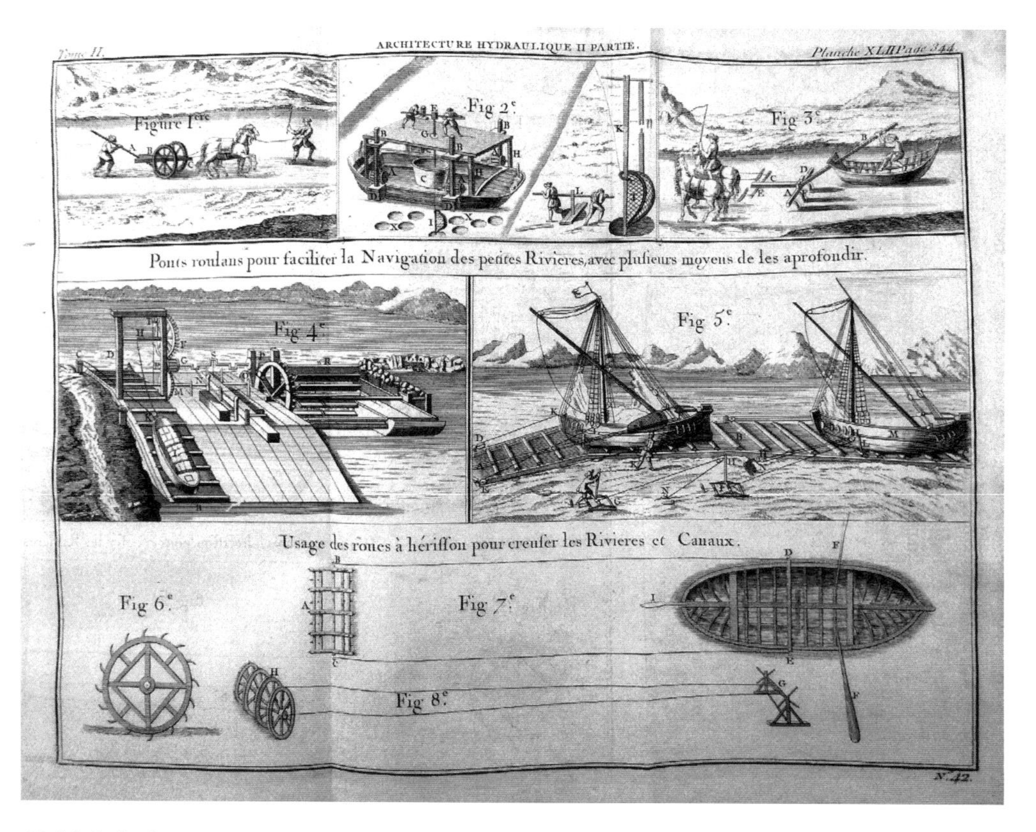

39. Methods of navigating a small river, dredging it and using a floating bridge, from 'Architecture Hydraulique' by Belidor

156 WYJS/CA MIC2/2 21st June 1759.
157 WYJS/CA MIC2/22 Oct 1761.
158 WYJS/CA MIC2/2 31st Jan 1759.
159 WYJS/CA MIC2/1 14th Feb 1758.
160 WYJS/CA MIC2/4 28th Feb 1761.

year. According to the National Archives, a general guide to the equivalent value of £1 in 1760 would be about £102 in 2017; it would be about 10 days wages for a skilled labourer. The total of these sums is £9887: 8s: 4?d; the National Archives currency converter shows a rough equivalent value in 2017 as being £1,013,087.88. Question marks in the table below indicate where individual numbers were not clear in the original record.

Estimate to Dec 31st 1760

Obtaining the Act	£2075: 15s: 2d
Timber	£1141: 7s: 11?d
Iron	£155: 3s: 1d
Stone	£540: 7s: 0d
Land	£732: 15s: 0d
Cement	£69: 3s: 4d
Digging	£617: 1s: 0d
Freight	£65: 16s: 1d
Brick	£37: 16s: 0d
Salaries	£435: 5s: 4d
Stationaries	£13: 2s: 2d
Wages	£1192: 0s: 4?d
Interest	£196: 8s: 9d
Utensils	£88: 12s: 6?d
Pettes	£142: 17s: 5?d
Incidents per Comers.	£58: 10s: 11d
Pitch	£14: 0s: 0d
Fodder	£4: 4s: 0d
Cordage	£20: 0s: 0d
Cash in the clerk's hands	£69: 0s: 10?d
Cash in Mr Nickalls hands	£71: 17s: 8?d
Cash in Comers. hands viz Treasurer	£1266: 3s: 8?d
Cash paid since to Feb 28th	£880: 0s: 0d

Buying Land

From the beginning of his involvement with the scheme Smeaton had seen potential difficulties over land purchases and emphasised the necessity of dealing with land and millowners with consideration. On 14th August 1757 he wrote: 'I am aware that in a work of this extent many person's property may suffer alteration; it may be of use to know which particular parts are likely to meet with the Greatest Difficulty from the Proprietors thereof.'[161] He was also aware that apart from land purchases, issues like access to land, provision of bridges and compensation for damage on private or public roads, as well as interruption of the business of the mills would arise. A few owners, like Sir George Savile with land in Thornhill, would be keen to agree a price as they supported the navigation, but others, like the owners of the mills who opposed it, could hinder the process, and yet others would haggle to get the best price. In the case of an owner refusing to treat for thirty days, the Parliamentary Act allowed for a jury of twelve men to be called to decide on the issue, but it was clear that this would be a last resort. However there was at least one significant long-term grievance case brought against the navigation, that of William Banks, a Leeds merchant who owned Sands Mill at Dewsbury[162]. He had been very much opposed to the navigation but as a result of the White Hart meeting he had been promised that the navigation would buy his mill and that he could become a proprietor of the navigation. Unfortunately the change from proprietors to subscribers made during the bill's passage through Parliament in 1758 meant this agreement was impossible to honour until a second bill in 1769 when a Company was formed. Banks claimed and won retrospective damages from the Calder and Hebble Navigation Company at a case held at York Assizes before a jury in 1774.[163]

A List of Owners of Mills within the intended Navigation.	
Wakefield and Horbury Mills	Sr Lionl Pilkington Bart.
Dewsbury new, Mill	Mr Banks
Dewsbury Mills	Mr Greenwood
Mirfield low Mill	Mrs Witton of Burfit
Ledger and Batty's Mills	Sr John Kay Bart.
Halls Mill	Mr Darby
Kirklees and Brighouse low Mills	Sr John Armitage Bart.
Brighouse upper Mills	Thomas Thornhill Esqr.
Ealand Mills	Sr Geo. Savile Bart.
Bankhouse Mill upon the Brook	a Charity

40. Panel listing the owners of mills from John Smeaton's Plan

By 11th January 1760 Smeaton announced that the project was ready for purchases of land. The first three cuts at Wakefield, Lupset, and Horbury had been staked out and he had measured the land required from each proprietor. The plans of the projected cuts were sent to the commissioners with the quantities of materials required. Wakefield was to

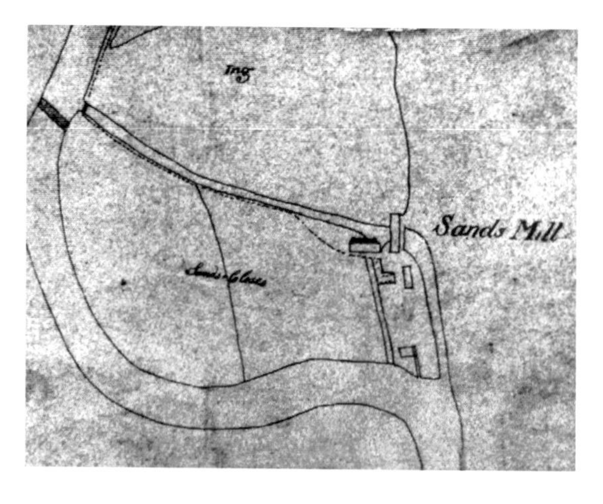

41. Sands Mill from 'A Map of the Mannour or Rectory of Dewsbury, surveyed by Jno Parsons and Jno Thompson' 1761.

161 WYJS/CA MIC2/1 14th Aug 1757.

162 On Smeaton's plan this is referred to as Dewsbury New Mill.

163 HAS/T 1922 *Our Local Canals* Chas. Clegg.

be where the work began and so they had 'examined again in Wakefield the conveniences of a piece of Ground ... a Close of about an acre and a half bare Measure, lately purchased by Mr James Milnes and through part of which our intended Cutt is to pass.' This was the first piece of land purchased for the navigation on the 11th February and Nickalls was given the key on 16th February. It was to include the works yard which is described in operation on Monday 26th May at the start of the Journal. The commissioners paid £150, the price Mr Milnes had recently given for it, 'and they allowed £21 more 'for the work lately laid on for the Repairs of the Fences and for the Writings'.[164] Negotiating with the owners of land and mills was a continuing process as the work advanced upstream and agreements were often reached a matter of weeks or even days before the start of work.

Also in January 1760 Mr Wilcock of Thornhill was invited by the commissioners to become their leading land valuer. During the subsequent period before Smeaton arrived the purchase of land was the main subject of the intense correspondence between Smeaton, Wilcock, Simpson and Nickalls as they reported on meetings, visits and conversations: the gradual negotiation of land prices was clearly much debated by all interested parties. On 13th February Mr Wilcock valued the land required for the first three cuts at £80 an acre for the Wakefield Cut, £60 for Thornes and £40 for Horbury. Most of the proprietors

A List of the principal Owners of Lands contiguous to the River, as more particularly expressed in the large Map.

from	North Side	from	South Side
1 to 2	Mr. Ramsden & various Owners	1 to 2	Mr. Ramsden and various
2 to 3	Mrs. Witton	2 to 3	Mr. Beatson and various
3 to 4	Horbury comn. Lands & various	3 to 4	Geo. Fox Lane Esqr.
4 to 5	Mr. Edw. Oats and various	4 to 5	Sr. Geo. Savile Bart.
5 to 6	Osset comn. Lands and various	5 to 6	Mrs. Thornhill and various
6 to 7	Mr. Ward and various	6 to 7	Mr. Turner and various
7 to 8	Mrs. Witton of Burfit & various	7 to 8	Sr. John Kay Bart.
8 to 9	Richd. Beamont Esqr. & vars.	8 to 9	Richd. Beaumont Esqr.
9 to 10	Sr. John Armitage Bart.	9 to 10	Sr. Lionl. Pilkington Bart.
10 to 11	Josha. Horton Esqr. & various	10 to 11	Thos. Thornhill Esqr. & varius
11 to 12	Sr. Geo. Savile Bart.	11 to 12	Sr. Geo. Savile Bart.
12 to 2	Willm. Gream Esqr.	12 to P	Charity Lands & various

42. Panel listing the principal owners of land from John Smeaton's Plan

164 WYJS/CA MIC2/4 11th Jan, 11th Feb, 16th Feb 1760.

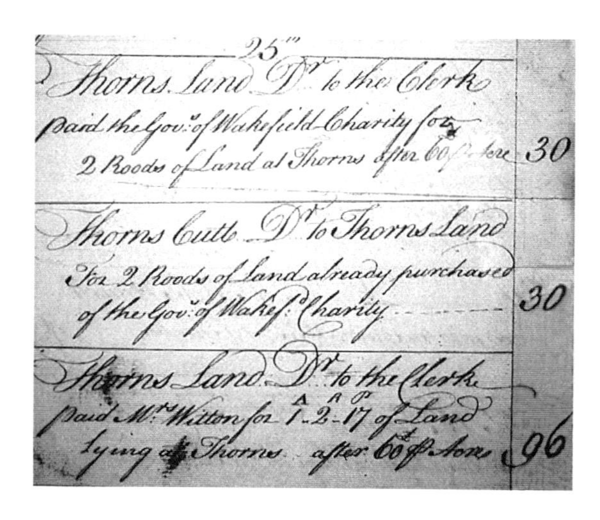
25th
Thorns Land Dr. to the Clerk
Paid the Gov.rs of Wakefield Charity for
2 Roods of Land at Thorns after 60 p. Acre 30
Thorns Cutt Dr. to Thorns Land
For 2 Roods of Land already purchased
of the Gov.rs of Wakef.d Charity ---- 30
Thorns Land Dr. to the Clerk
A R P
paid Mrs Witton for 1-2-17 of Land
lying at Thorns after 60£ p. Acre 96

43. Payments for land at Thornes, 25th April 1760

in the area of Thornes and Lupset, including Mrs Witton of Lupset Hall, agreed to take £60 per acre fairly quickly. Work was to be started there and purchases were completed by 28th April following small concessions to two reluctant owners, the Governors of Wakefield Charities and Mr George Charnock. In some cases the quantity of land was fairly small and thus the amounts of money were low, for example Charnock was eventually paid £4 17s 6d, but setting precedents in individual negotiations was a concern for Simpson and the commissioners.[165]

On 23rd February Wilcock reported that 'the Principal Inhabitants in Wakefield acquainted us that they expected £124 an acre for the Wakefield Cutt' based on their calculation of yearly rents. Land near Wakefield was particularly valuable because of the busy market town's need for grazing and market gardening, but in addition there is no doubt that this extension to the navigation was still unwelcome to some in the town. Smeaton wrote from London 'I am not surprised at the Wakefield People asking an exorbitant price but apprehend if something more than the value be tendered, and in Consequence, that the Charge of a Jury fall upon themselves ... I don't doubt but they will begin to think of more moderate terms.' Smeaton was well aware of this continued opposition commenting in a letter from Austhorpe dated 30th June 1760 after rain had caused problems on the works that 'I found the Wakefielders had been very Industrious in propagating a report concerning the Greatness of our Damage and that all our Tools were carried away etc from whence we may judge what they wish to be true.[166]

Whilst he was still in London Smeaton was given the task of calling on the several gentlemen landowners in this area who were up in town: Sir George Dalston, Mr Fox Lane Esq. of Bramham, Lord Strafford and Captain Ramsden. It was difficult to reach agreements with such men as they delayed their decisions, keeping the business at arm's length. Smeaton had visited them all before 25th March but only received a reply from Ramsden whose land at Lupset was needed to start the work; the others referred the

165 WYJS/CA MIC2/4 Jan 1760, WYJS/CA MC2/22 13th Feb, 28th April 1760.
166 WYJS/CA MIC2/4 23rd Feb, 1st March, 30th June 1760.

matter to their agents in the country. As the time to begin work drew near the threat of the jury and the costs this would incur were eventually used to apply pressure. In Wakefield the commissioners increased their offer to £110 for Wakefield land based on the price per acre already given for James Milnes' land. By April 29th the Governors of Wakefield Charities eventually agreed to accept this. One or two reluctant owners like Mr Shepherd who had a garden near Fall Ing accepted 'the same Terms as the rest of the Proprietors in the Wakefield Cutt', as did the Earl of Strafford who agreed the sale of his land there on 22nd May.[167]

It is agreed between the Governors of the free Grammar School of Queen Elizabeth at Wakefield and Mr. Joseph Nickalls, Mr. Thomas Simpson and Mr. William Wilcock, who treat on Behalf of the Undertakers for extending the Navigation upon the River Calder from Wakefield to or near to Sowerby Bridge that so much of the Close in the Occupation of Daniel Shillitoe at Wakefield as shall be covered by the Bank on the North West Side of the Cut to be made therein, as the Slope next the Cut and the haling Path on the Top of the said Bank to the Commencement of the Slope on the Land Side, will take up, shall be purchased by the said Commissioners or Undertakers after the Rate of one hundred and ten Pounds an Acre as well as the Ground which shall be used in the Cut and the South East Bank thereof as far as the Hedge dividing the said Close from a Garden in the Occupation of James Brooksbank, and that the Ground taken up by the Slope to the Land Side of the said North West Bank shall be estimated in Damages according as that Slope shall be adjudged to affect and lessen the Value of the Ground in Prejudice of the Owners and Occupier, and shall be made Satisfaction for after the same Rate.

Witness,

W. Chippindall.

Jos.h Nickalls

Tho.s Simpson

44. Agreement with the Grammar School Governors. 25 April 1760

167 WYJS/CA MIC2/4 26th Feb, 29th April, 25th March, 22nd May, 1760.

Smeaton met Mr Lane (Fox Lane) who owned land in Netherton on the 22nd May in London. Smeaton was about to return to Wakefield and a decision was becoming essential, so he warned that the matter would be referred to a jury if Lane did not also accept the offer. Smeaton reported Mr Lane's strong reaction: 'he should take it much amiss and … if they did this they might expect all the obstructions that he could give'. In March the Horbury people had demanded £60 for their pastures which were regarded as similar to Mr Lane's land in the parish of Netherton and therefore should be acquired at the same price. However by 20th June all the Horbury Pasture wanted for Horbury Cut had been purchased for £52 10s an acre from the 'Inhabitants of Horbury' as agreed in a parish vestry meeting. Mr Lane was still holding out for £60 an acre despite Simpson's further threat of an adjudication in a letter to his agent in June 1760. It was not until November that his steward, Mr Heelis, wrote 'the Gentn need not give themselves any further Trouble regarding the Jury provided for to be summoned, for the Squire will certainly accept of the Sum offer'd by them.'[168] At the beginning of the following year Washingstone and Horbury Bridge land was required. Again the Horbury Vestry meeting asked for £60. Nickalls discussed this purchase with the Constable of Horbury, Robert Carr, who was influential in the Vestry decision to accept £55 per acre. Mr Lane had other land required for Washingstone Lock and Cut which he agreed to sell at fifty guineas an acre.

However in January 1761 there appeared to be opposition to the navigation in some quarters. Simpson reported to Nickalls that wheelbarrows at the local quarries had been broken by 'evil disposed Persons' and he suggested that the commissioners should prepare notices 'at Horbury and other church doors etc. setting forth the nature of the law … and signifying a reward.'[169]

Late in December 1760 Smeaton and Nickalls surveyed the area for the next section of the work towards Dewsbury. They were considering the practicality of a change in the original route designed for the navigation by constructing a long canal from the Figure of Three Lock into the river on the opposite bank to Dewsbury Church. Smeaton recorded in the Journal the advantages of this potential alteration which became known as the long cut.[170] On 3rd March 1761 he wrote to the commissioners, 'I have thought of the proposal I made to the Commissioners over and over and in every view think it eligible to execute the long cut'[171]. By August 1761 negotiations for land in Thornhill required for the cut were taking place. The main landowner here was Sir George Savile whose tenant, Mr Wilcock, the commissioner's valuer, leased property near Dewsbury on the estate. Both were, of course, supporters of the navigation. A Mr Taylor and Mr Horsfall were asked to do valuations for this area to avoid a conflict of interest. Unfortunately after late 1761

168 WYJS/CA MIC2/4 12th May, 22nd May, 5th June, 10th Nov 1760.

169 WYJS/CA MIC2/4 19th Jan 1761.

170 WYJS/CA MIC2/16 20th Dec 1760.

171 WYJS/CA MIC2/4, 3rd March 1761.

the copying of letters appeared to become spasmodic so that further land purchases are hard to identify from this source and the accounts do not include payments for land by this time.[172]

As in this case, it is likely that the difficulties encountered at Wakefield in negotiating the price of land may not have been replicated later. In most cases the owners upstream recognised the advantages the waterway would bring them. Advertisements in the local papers for property near the river quickly included a reference to the navigation. For example as early as 29th January 1760 Blake Hall in Mirfield was advertised to be let on the death of Mr Turner, the owner. It was confidently described as 'Pleasantly situated near the Church in Mirfield, adjoining the Turnpike Road now making from Dewsbury to Elland, and not above quarter of a mile from the River Calder, now making navigable from Wakefield to Elland, and up to Sowerby Bridge.' On 8th June 1762 Lupset Hall was advertised for sale as 'lying upon the River Calder now navigable above Horbury'[173]

Gradually sections of the works yard activity began to follow the work up the river. On 3rd June 1762 half of the smith's shop went to Dewsbury and the minion mill to

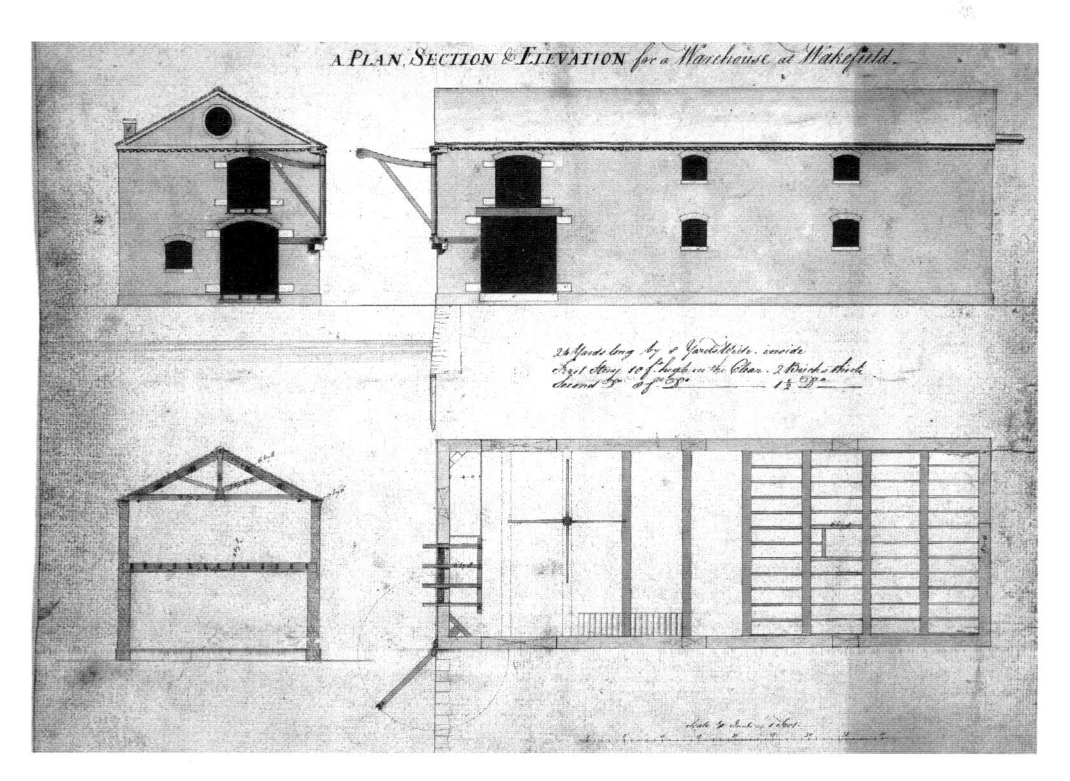

45. Smeaton's design: A plan, section and elevation for a warehouse at Wakefield

172 WYJS/CA MIC2/4 8th Feb, 23rd March, 24th Aug 1760

173 LI 29th Jan 1760, LI 8th June 1762.

Horbury,[174] and on 17th June 1762 the commissioners began to look for land in Dewsbury for a warehouse.[175] This had not been settled by May 1763 when the commissioners were again looking for riverside land in Dewsbury for a temporary warehouse. However in late June the sheds from Wakefield Yard were taken down and carried up to Dewsbury in order to be erected as temporary premises in Mr Wilcock's orchards. On 23rd June they ordered that a warehouse be built of brick in the Wakefield store yard 24 yds by 8 yds within and 2 stories high, and a crane was added in late July.[176] The busy Wakefield works yard was being transformed into a warehousing facility as goods started to move up and down the first section of the new navigation. The large-scale construction activity on the Wakefield riverside and the considerable numbers of labourers accommodated in the town and area were moving on up the river.

The workers and working arrangements

Although Smeaton had been involved in surveying the River Clyde and River Wear he had not supervised the building of a river navigation. For the Halifax gentlemen too, it was a major venture with unknown risks. The work described in the Journal was undertaken by a wide range of men, from labourers to experienced engineers and craftsmen. Inevitably this work force and many aspects of the working arrangements and practices were evolving as the labour requirements and the specific demands of this navigation became clearer. As time went on the pattern of employment changed from directly-employed workers at all levels to contract work led by proven individuals. The roles within Smeaton's initial team of Nickalls, Gwyn and Wilson also changed.

Over the 1760 and 1761 seasons the payment of the wages of the hands in the yard and at the other sites was the responsibility of Joseph Nickalls. On 5th March 1760 he wrote to Simpson: 'workmen's and other expenses will increase Daily shall soon be in want of Money'; this was the first record of his weekly requests for money from Halifax. From before the beginning of the Journal in late May 1760 it is clear that a considerable number of hands and labourers had already been employed. In the accounts of 24th February the first set of wages in the storeyard amounted to 19s: 31/2d, but by 18th May, just before Smeaton arrived, the workforce had multiplied so that Nickalls was paying out £11: 15s in the storeyard, distributing wages at Thornes Cutt amounting to £8: 15s: 7d for digging and 18s: 9d for 'grubbing roots', and also supplying Horbury workers with 10s for unspecified work. By the end of the year there was a total wage bill for these groups amounting to around £36 per week.[177]

---

174 WYJS/CA MIC 2/16

175 WYJS/CA MIC 2/1 17th June 1762.

176 WYJS/CA MIC 2/2 9th - 15th May, 27th - 3rd July 1763.

177 WYJS/CA MIC2/22 1760.

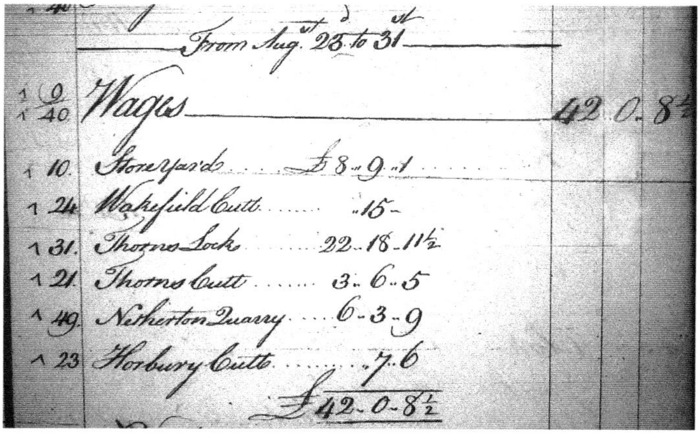
From Aug. 23 to 31

40	Wages	42 0 8½
10	Storeyard	£8..9..1
24	Wakefield Cutt	..15..
31	Thorns Lock	22..18..11½
21	Thorns Cutt	3..6..5
49	Netherton Quarry	6..3..9
23	Horbury Cutt	..7..6
		£42..0..8½

46. Wages paid 23rd to 31st August 1760

Little is known about these unnamed labourers referred to by Smeaton early in the Journal as 'the Commissioners' men' or 'our men' who were directly employed at this time. Their work is recorded in the Journal as they were occupied in a variety of jobs in the yard, prepared the sites for the locks, or began the digging or masons' work. At first it appeared that many of these labourers and craftsmen were local and there were problems in keeping them on the job. In July the lack of masons was an issue and they were encouraged to sign up by the promise of work over the winter. In August some of the hands disappeared to the harvest. By December 1760 the Journal records the piecework rates for masons working at several nearby quarries. The conditions of the work the men were expected to do is revealed by the commissioners' resolution on 18th June 1761 that those employed under the direction of Nickalls and Smeaton should only work a six-day week and a 12-hour-day 'except in cases of emergency'.

The first major 'Undertaker' or contractor, who brought a team of his own men, was Jeremiah Platts from Lancashire where a number of navigations had been constructed in the preceding years. He was one of at least three applicants to the commissioners 'to cut the river'.[178] On 23rd March 1760 they requested a reference for him from a Mr Lee or Legh of Wigan who stated that: 'Jeremiah Platts whom you enquire after has been Employ'd by me to make two Canals which he performed reasonable and well, he was employ'd in making the Sankey Navigable, not as the first Hand but as a Person in whom much Confidence was put, and in his Pursuit at the Work I do believe gave full satisfaction.' Platts sent a tender for digging in response to which Smeaton, with characteristic common sense and caution, wrote to Nickalls that an offer significantly above or under a general price makes him suspicious, which he thought was the case here; but 'by being so much under … all Other Prices, … tis much easier to rise than to fall Prices agreed on.' Smeaton requested that Platts answer several questions about how he would dig out the banks and deal with the material which had been displaced, about the laying of turf and ramming of

178 WYJS/CA MIC 2/4 23rd March 1760.

the banks, and the completion of both locks and cuts. Here Smeaton was indicating the standards that he expected in completing this contract work, and indeed, not long after Platts' arrival, Smeaton 'was oblig'd to expostulate with Mr Platts' who 'tho' he goes with his Work very readily, yet seems more inclined to get the matter out of the Cutt than to consult our Convenience or Orders in laying it.'[179]

On April 29th 1760 Simpson asked Platts to come over with his men, agreeing at the same time that barrows and deal planks would be supplied. 'You'll be wanted immediately therefore the utmost despatch is required' he emphasised. Platts was asked to arrange the accommodation for himself and his men, presumably in Wakefield or nearby, and to 'get your Edge Tools ready'. The agreement recorded at a commissioners' meeting on 28th May 1760 was that Platts would be paid weekly for his whole team at the rate of 'one shilling per Man till he has finished some part of the cut, which when measured he is to be paid for what is further due to him or agreeable to his contract.' A labourer's average wage on a local turnpike at that time was just over a shilling a day[180] so this arrangement must have encouraged some very hard work. By late May Platt's team had already been working on the Thornes to Lupset cut and had begun the lockpit at Thornes. These were the 'Undertaker's men' referred to by Smeaton in 1760. It seemed that the work began to meet Smeaton's requirements as his later journal entries suggest that Mr Platts was a trusted member of the team.

More contracts were offered after the first part of the works at Wakefield, Thornes and Lupset was completed. In June 1761 the commissioners agreed that Smeaton and Nickalls should contract with someone for the Masonry of Horbury and Washingstone Locks, ' at a price not exceeding 3s per solid yd'. By July 1761 several contractors had made bids ... Those who are disposed to engage for that are Wm Charnley & Co and Josh Riley & Co they are all sound Workmen ... ' On August 20th Smeaton sent a letter with William Charnley to Halifax saying that 'the latter had completed Washingstone Lock and was well advanced on the Dam ... He had got together a Number of Able Hands which he shall be oblig'd to discharge when his present work is completed, unless he can enter into a further contract with the Committee.' There were plans for six locks during the 1762 season for which large quantities of materials would be needed so that if the commissioners were to give Charnley the contract for two locks he could then employ his men in the quarries at the end of his present work. Smeaton wrote of Charnley 'the work he has done by contract has been equally well performed with what was done by Day under him as foreman and the whole executed in a Substantial and Satisfactory Manner ... whatever Contracts he makes with the commissioners will be faithfully perform'd.' Smeaton suggested in this letter that

---

179 WYJS/CA MIC 2/4 8th June 1760.

180 On the Wakefield to Weeland Turnpike in 1761 labourer's pay was from 1s to 1s 2d. a day. (Ref: Albert William, *The Turnpike Trust Road System in England*, 1663-1840, p 162)

taking on a contract for work on a lock, should, during the following year, include 'getting stone, hewing, leading,[181] putting together, Drainage of the water and in short everything after the Foundation is piled.' In this way the whole process of getting and working stone and all the building work in stone became the contractors' responsibility: the payment of individuals or small teams of masons on piecework began to disappear from the accounts.

During the winter of 1761/2, following the dismissal of Nickalls, the awarding of new contracts was to be decided at the commissioners' meetings with Smeaton advising. The minutes include many more direct orders about what work was to be done. Smeaton attended and doubtless advised a meeting of 2nd December 1761 which announced that in sites with soft foundations the use of stone for stern sills and recesses should be replaced by wood, that there would be a lock upon the west end of the Long Cut opposite to Dewsbury, and that the tailgates of all the locks were to be made double. More contractors began to be employed in more areas of work. Carpentry was to be let by measure and more contracts for completing the masonry at locks and for building bridges awarded. By December 16th 1761 responses had come from Joshua Wilson, John Longbottom,[182] Edward Sykes and William Charnley for building locks and from Luke Holt and John Topham for carpentry. In January 1762 Mr Platt's proposal to build drains was agreed by the commissioners, and John Cousin's bid for the execution of the iron work by the foot was accepted. Most of these men had already gained experience of what was required during the previous two seasons.[183]

Perhaps some of the commissioners felt that there was, by then, a team who had gained sufficient experience and could take the work forward without the role of a Deputy Surveyor. Whatever their reasoning they took the decision not to appoint a direct replacement for Nickalls without, apparently, consulting Smeaton. In March 1762 writing from London to Simpson, Smeaton expressed a hope that a new Deputy Surveyor would be appointed soon to deal with a problem which had arisen. However the commissioners were already advertising two roles of 'Superintendent of Masonry and Digging' and 'Superintendent of Carpentry'. On 17th March John Gwyn was appointed to oversee the carpentry and smiths' work at £50 a year and on 31st March Mathias Scott was appointed to superintend the masonry and digging at £60 a year, salaries which would be paid quarterly. The resolutions made by the commissioners during the winter of 1761/2 seem to have been designed to indicate to Smeaton that they wanted to have the final word, not only about their employees but also about the work to be done. Smeaton may have been feeling rather beleaguered: he had lost his close colleague, Nickalls, and now he had lost the role of Deputy, a post for which he had argued against much opposition at the outset of the project. The change

181 leading - bringing materials by cart.

182 This may be the John Longbotham who became chief engineer for part of the Leeds & Liverpool Canal.

183 WYJS/CA MIC2/2 13th Jan 1762.

CALDER NAVIGATION.
WANTED,
TWO Surveyors for the ſaid Navigation, one to ſuperintend the Maſonry and Digging, the other the Carpentry, both under the Direction of Mr. *Smeaton.*
Such Perſons properly qualified and well recommended, on applying to the Commiſſioners of the ſaid Navigation the 7th of *March* next, at the *Talbot*, in *Halifax*, will have due Attention and Encouragement given them.

47. Advertisement in the Manchester Mercury, 23rd February 1762

CALDER NAVIGATION.
This is to give Notice,
THAT if any Perſon is willing to contract for the Drainage of *Cowper Bridge Lock, Kirklees Mill Lock, Anchor Pit Lock, Brighouſe Lock,* and *Lillands Lock,* or any of them, that the Commiſſioners will be ready to treat with him on *Thurſday* the 23d of *December* 1762, at the *Talbot*, in *Halifax.*

48. Advertisement in the Manchester Mercury, 14th December 1762

of terminology from 'deputy surveyor' to 'resident engineer' seems to better describe the working relationship between the leading engineer and his deputy. Smeaton was later to stress the importance of a good relationship between the resident engineer and the committee directing him,[184] something which seems to have been lacking in Nickalls' case, perhaps because of the distances in the first seasons of work - a good half day's ride between Halifax and the Wakefield yard in 1760 and 1761.

However the use of contractors meant that the burden of dealing with a great many suppliers of materials as well as the calculations of pay for many employees and piece workers, which had so added to Nickalls' worries was largely removed from his two successors by the spring of 1762. In the commissioners' minutes of December of 1762 and 1763 the same few names: Platts, Charnley, Wilson, Luke Holt and one or two others appeared. They proposed prices for work on listed locks, dams and bridges working steadily up the Calder and were paid substantial sums regularly. The Clerk, Simpson, was to be paid £10 for extra service this year and an extra £20 a year in the future, suggesting that it would be he who would deal directly with contracts and contractors. The workers were the responsibility of the contractors whose problems remain unrecorded, although in a letter of February 1763 Smeaton was requesting that land purchases at Kirklees Mill be made quickly as `Mr Platts has got 130 hands upon Batty Mill cut' who would soon require further employment. The difficulties arising from the management of such numbers can only be imagined.

In their dealings with the contractors who worked on the navigation Nickalls and Smeaton expected hard work but were supportive when this was done well. They recognised the value of experience and the names of those working on the navigation tended to remain constant, in many cases with more responsibility being given over time. This was

184 Skempton, Ch X, Denis Smith, 'Professional Practice', London 1981, p 225.

clearly in their interests: 'It appears to Mr Smeaton and Self, that the most likely way to have the whole perform'd, will be to Let a person be continued in the same Kind of Work as much as may be consistant for his Performance within a desirable Time, by this any One will become the more versd.'[185] Smeaton was later to advise that 'it was eligible to do all the work you possibly can by contract', as long as the lowest price was not the first consideration.[186]

A number of contractors from the Calder scheme continued to forge busy careers during the great canal-building era of the late eighteenth century, indicating a rich legacy of expertise developed working under Smeaton's guidance.[187] Their names appeared in later schemes, many of which were also supervised by Smeaton. Luke Holt from Middlestown, employed from the start as a carpenter and later as the carpentry contractor, was from 1769 to 1772 a joint resident engineer on the final section of the Calder navigation up to Sowerby Bridge, working with Robert Carr, the mason and Horbury vestry spokesman. Holt then surveyed a new link from Huddersfield to the Calder and Hebble, to be known as the Sir John Ramsden Canal, for which an Act was passed in 1774.[188] Later in the 1770s he was resident engineer working on the Old Dock at Hull, recommended for the post by Smeaton, and subsequently took the same role on the Weighton drainage and navigation scheme. He did further work on the Calder at Ossett Mill and Ledger Bridge towards the end of the century.[189] Joshua Wilson had resigned as Master Mason in 1761 but appeared as a major contractor soon afterwards building several locks and dams and also Elland Bridge. In 1767 he was acting as masonry contractor on the Ure Navigation and Ripon Canal.[190] William Charnley, who had moved from a foreman to a main contractor on the Calder, was awarded an increasing number of contracts for work during the period of the Journal. Later Smeaton sent for Charnley to open up and work quarries at the start of his next major waterway, the Forth and Clyde Canal and by 1769 Charnley was supervising the masonry on the first lock there with John Gwyn supervising the carpentry. Gwyn had been given more responsibility whilst working on the Calder and when the navigation opened to Dewsbury he was also asked to take 'account of the Loading of all Boats going up and down the river and receive all the Tolls'.[191] Smeaton recommended him to others or used him as his resident engineer on many further occasions into the

185 WYJS/CA MIC 2/4 2nd July 1761

186 Skempton, Ch X, Denis Smith, 'Professional Practice', p 225, London 1981.

187 Mike Clarke notes that the Halifax areas seems to have been a centre for canal expertise with a number of local men becoming involved in later schemes like the Leeds & Liverpool Canal. (Ref: '*Understanding early English canal technology - an Austrian view*')

188 Charles Hadfield *Canals of Yorkshire and North East England Vol 1*, Plymouth 1972, p 61.

189 Ed Skempton and Grimes *A Biographical Dictionary of Civil engineers in Great Britain and Ireland, Vol 1* London 2002, pp 333-334.

190 Charles Hadfield, *Canals of Yorkshire and North East England Vol 1*, Plymouth 1972, p 111.

191 WYJS/CA MIC2/2 4th Nov 1762.

late 1780s, describing him as 'the very best man for executing orders I have ever met with' in April 1767. In the winter of 1768 he was warning some new employers in Hull not to judge on first impressions as Gwyn wrote and spoke 'the London Vulgar...with a knack of misapplying an heap of Fine Words and Phrases'[192] but this did not reflect his engineering abilities.

192 Ed Skempton and Grimes, *A Biographical Dictionary of Civil Engineers in Great Britain and Ireland, Vol 1* London 2002, pp 286-288.

CHAPTER 4

AFTER THE JOURNAL

This study has followed the history of the Upper Calder Navigation from the origins of the scheme to the end of the fourth season of work in 1763 after which Smeaton's Journal of the work is not in the records or may not have continued.

From 1762 there was a small toll income from goods travelling above Wakefield to Dewsbury. However the spending by the end of 1763 amounted to £36,000, the sum Smeaton had estimated for the whole work.[193] A further £20,000 was raised and the navigation was open to Brighouse by the end of 1764. Although Smeaton felt that progress was good, some of the subscribers were restive. Many of those from the Halifax area were reluctant to continue to finance the project beyond Brooksmouth.[194] When a new subscription was proposed the meeting was held in Rochdale and a new committee based there was elected to oversee the work. They immediately wrote to James Brindley, famous for his role in the recent construction of the highly successful Bridgewater Canal. When Brindley came across to look at the Calder project he was appointed 'Surveyor, Manager and Undertaker' whilst Smeaton, Gwyn and Scott were discharged. By the end of 1765 the navigation was open to Brooksmouth but in June 1766 it appeared that Brindley had moved on to other work including planning a canal to Rochdale. Simpson was left in charge and the work continued.[195]

However further extensive damage to the navigation occurred in the late 1760s. Following some very heavy flooding George Savile wrote despairingly to Lord Rockingham on 24th October 1767, 'I have just got back from looking at some of the damages on the Calder which for the present have rendered the navigation above Wakefield null and void.

---

193 Charles Hadfield, *The Canals of Yorkshire and NorthEast England*, Vol 1, Newton Abbott, 1972, p. 48.

194 LI 18th June 1785, Letter from an unnamed Commissioner, not involved previously in the scheme.

195 Skempton, Ch V, 'Rivers and Canals', Charles Hadfield, London, 1981, p 108.

ANNO REGNI

GEORGII III.

REGIS

Magnæ Britanniæ, Franciæ, & Hiberniæ,

NONO.

At the Parliament begun and holden at *Westminster*, the Tenth Day of *May*, *Anno Dom.* 1768, in the Eighth Year of the Reign of our Sovereign Lord GEORGE the Third, by the Grace of God, of *Great Britain, France*, and *Ireland*, King, Defender of the Faith, &c.

And from thence continued, by several Prorogations, to the Eighth Day of *November* 1768; being the Second Session of the Thirteenth Parliament of *Great Britain*.

DIEU ET MON DROIT

LONDON:
Printed by GEORGE EYRE and ANDREW STRAHAN, Printers to the King's most Excellent Majesty. 1824.

49. First page of printed Act of Parliament 1769 9 Geo lll c71

I have no guess at the damages. They are however probably considerably more than the money we have left and nobody seems dispos'd to advance any more. In the meantime we are losing about £50..£60.. £70 per week on Tolls. What we want is somebody to lend us about £10,000 at 5 per cent but without a prospect of ever seeing principal or interest.'[196] In February of the next year more damage was inflicted. Smeaton agreed to return to assess what was required and recommended repairs to the value of £3,000, but even after much of this work had been done the navigation was again closed by flood damage. At this point a new Act of Parliament was sought, as ever supported by Parliamentary evidence from Smeaton. As a result the Calder and Hebble Company was created with subscribers becoming proprietors owning shares. From May 1769 Simpson worked with Luke Holt and Robert Carr as joint superintendents, and in September 1770 the waterway was open to Sowerby Bridge. At last a firework celebration was sponsored by some of the gentlemen and the navigation began to repay its investors with trade growing quickly and dividends increasing.

The Act of 1768 included an outline of the tolls which could be taken for the carrying of different items on the navigation and the list shows the huge range of goods expected including: coal, cinders, lime, dung, wool, cord, vegetables, and heavy goods such as iron, stone and flags, tiles, hay and timber, flour, and butter, and many other grocery items, cloth and bales measured by the ton.[197] For the population of the upper Calder valley the navigation provided improved access to basic raw materials for their economy and their households, including wool for the clothiers, coal for their fires, and lime for the soil. In addition the burgeoning import trade brought an increasing range of new foods and fashionable or exotic items. There was easier distribution for the area's cloth manufactures and other products from the region, especially the fine quality flagstones and setts of Elland Edge.

196 Sheffield City Archives, WWM/R/1/864, Rockingham Papers.
197 WYJS/WA QD5/4/6, Act of Parliament, 1768.

There were many further alterations to the route and repairs and improvements required to the structures, but the navigation continued to be a success. Inevitably this opened up prospects for pushing other water routes across the Pennines and also creating a number of new canal or navigation links between other centres in the West Riding. In particular the Leeds & Liverpool Canal was begun in 1770. In 1794 a Rochdale Canal Bill was passed to link Sowerby Bridge to Manchester, and was completed by 1799. Remarkably the final 1.75 mile link up the River Hebble into the centre of Halifax, proposed in the 1758 Calder Navigation Act, was not completed until the mid-1820s.

HALIFAX JUBILEE.

ON Wednefday the 17th of July Inftant, will be held, A JUBILEE, at the New Warehoufe near *Sowerby-Bridge*, on account of the CALDER NAVIGATION being opened; where for the Entertainment of the Company will be a

Concert of MUSIC, upon the Water,
By the beft Band that can be procured from York, Manchefter, *&c.* and in the Evening, a Grand Exhibition of FIRE-WORKS, and a BALL.

The Ladies will be admitted gratis, and Tickets for the Gentlemen, at Half a Guinea each, may be had at the Bull's-Head in Manchefter; the Roebuck in Rochdale; the Talbot in Halifax; the Sun in Bradford; the Old King's-Arms in Leeds; and at the White-Hart in Wakefield

☞ In an Advertifement inferted in this Paper laft Week, the Word COMPLEATED was thro' Miftake us'd inftead of OPEN'D.——And it having been objected that the Advertifement laft Week wou'd be liable to be underftood as if the Company of Proprietors of the Calder Navigation, or their Committee, were concerned in this Affair; the public are hereby inform'd, that it is merely the Act of a few private Gentlemen, for the Entertainment of themfelves and their Friends.

☞ An Ordinary at the Talbot in Halifax.

50. Celebration of the opening of the navigation from the Leeds Intelligencer 9th July 1771

CHAPTER 5

CHALLENGES AND ACHIEVEMENTS

Smeaton's work on the Upper Calder Navigation does not seem to have received the acclaim of some of his other projects, particularly the Eddystone Lighthouse, perhaps because of the many setbacks and his replacement as Superintendent before the scheme was completed. However, as can be seen from the Journal, it was an extremely challenging project as his first navigation, and occupied him for considerable parts of the early years of his civil engineering career, during which time he tackled many problems which not only resulted in practical and organisational improvements but also in new scientific understanding. His experience of dealing with the Calder in flood when river banks burst, was, according to John Holmes, Smeaton's friend, put into use immediately when he was asked to combat the potential collapse of the old London Bridge in the early 1760s. Alterations had changed the course of the water. Smeaton advised 'throwing in a quantity of large ruff stones, which with the sand and other materials, washed down by the river filling up their intertices, would become a barrier to keep the river in its usual course.'[198]

Developments in the engineering of river navigations and canals were continual at this time, and Smeaton kept abreast of these in his reading, by visiting at home and abroad, and by sharing ideas with other engineers. He had promised that he was able to reduce leakage in locks and introduced the evidence of George Collett who described to the Commons Committee in 1758 the minimal amount of leakage of the gates at the Royal Naval Dockyard in Greenwich. To prevent leakage the lock gates needed to meet 'true to each other' and to a stone sill with good foundations.[199] Stone and brick-built locks needed different methods to support the walls compared to older locks which used timber backed by turf. Ground paddles, which allowed locks to be much deeper and filled more steadily,

198 John Holmes, *A Short Narrative of the Genius, Life and Works of the late Mr John Smeaton*, London 1793.

199 JHC, Vol XXVIII, p. 142.

had been used on the Brussels Canal, and Thomas Steers is reputed to have been the first person in the British Isles to employ them on three locks on the Newry Canal where he was the engineer from 1736 to when it was completed in 1741.[200] Smeaton's drawing of 'A Calder Lock' shows a ground paddle for the top gate, and gate paddles for the lower gates.[201] However the drawing for Cut Head Lock near Dewsbury appears to show gate paddles for both the upper and lower gates.[202] Dams too were designed carefully. In 1763 he allowed himself to reflect in his Journal when considering the dam at Horbury pasture: 'This Dam is in length near 180 Feet in some Places 9 F[ee]t water against it and the leakage thro[ugh]out the whole not half as much as cou'd be drawn by one hand Screw, and is a Master Piece of Work of its kind.'

He also pursued further theoretical research himself. Having recently reported on wind and water power, the work on the navigation gave him the opportunity to observe the power of men and horses. During the early lock pit excavations in 1760 he calculated the effectiveness of men in pumping water, and he designed an improved pump. He also calculated the standard for the power generated by a horse over an eight-hour day which was recognised as an accurate expression of 'horse-power' by later engineers.[203] Smeaton recognised the important role horses could play in pumping from the start, designing a horse frame to power water screws which was under construction at the Wakefield site at the beginning of the 1760 Journal. Two more were being made in the works yard on the first day of the 1761 Journal.

The hydraulic mortar using pozzolana and lime, which Smeaton had first used when he was building the Eddystone lighthouse, was again used for the Calder Navigation, proving its strength and durability. He experimented with a new source of hydraulic lime from Barrow to ensure it was equal to the Watchet lime used on the Eddystone. Describing this mortar, Smeaton claimed: 'This composition is of excellent use in jointing the stones that form the lodgement for the heels of dock gates and sluices, with their thresholds, &c. when of stone'.[204] Smeaton's mortar was an important step in the development of modern cement.[205]

However the Calder Navigation presented challenges in many other ways. From the start of his relationship with the commissioners Smeaton had been concerned about the effective management of such projects. He never allowed himself to be less than

---

200 Mike Clarke, *Understanding Early English Canal Technology - an Austrian view*, paper delivered at Birmingham 24 June 2017.

201 RS JS 6/18 General View of a Calder Lock - see illustration 54.

202 RS JS/6/32 Cutt Head Lock.

203 Skempton, Ch II 'Scientific Work' Norman Smith, London, 1981, p. 54.

204 Hurst, Rees, Orme, and Brown, *Reports of the Late John Smeaton* Vol 3, Longman, London, 1812.

205 A later adaptation of Smeaton's mortar was patented in 1824 by a Leeds man who set up work in Wakefield, Joseph Aspdin. His Portland cement, 'artificial stone', was a mix of limestone and clay heated and then cooled, ground and mixed with water.

courteous in his letters to the commissioners, although he became more frustrated by their interventions as time passed. The 1763 Journal[206] shows the errors that were appearing which Smeaton attributed to the pressure to speed up the work. He did not wish 'to lay blame on Particulars' but 'we have been continually press'd with such a Quantity of work going on at a time that 'tis with the greatest difficulty that anything can be got compleated in due time and as it shou'd be.' In October a postscript to Simpson read 'that the work be exposed to fewer accidents I must repeat my desire the Commissioners will not to be so pressing upon us for despatch: for by undertaking too much Work in a season little things will naturally be missed and upon small things great ones hinge'.[207]

Writing to his relative in London, John Holmes, in September 1764, he expressed even more disillusion: 'the works are going on very well but we are still subject to the same rubbs, that have ever tortured the undertaking from the beginning, and will continue till the End: that is; the difference of opinion, partys and disputes amongst our Rulers'[208] At this stage there were great differences between the Halifax commissioners and those from Rochdale over the financing of the extension to Sowerby Bridge which led to the transfer of meetings to Rochdale and Smeaton's own dismissal.[209] Later in his career when recording his ideas on construction supervision, his experiences on the Calder would surely have informed the view that 'the greatest difficulty is to keep Committees from either doing too little when any case of difficulty starts, and too much where there is none.'[210]

In letters to friends he was also critical of the commissioners' demands on his time. He complained to John Grundy, a fellow engineer, in a letter of 4th June 1764 'there is always some point of difficulty that needs particular attention, and when not the Trustees always think the business neglected if I am not there once a week!'[211] He had tried hard to introduce his preferred working structure, using a deputy or resident engineer to remain on site, and specialist foremen to oversee the different types of work. After the initial period of surveying, planning and sourcing materials his own role would be to draw up plans and design equipment as required, explain these to all relevant parties, attend meetings when available, approve appointments, respond to major problems and check and report on progress. He would not be involved with labour, payment for goods or wages, and above all he would not be available at all times in person, but acting in a supervisory role, keeping control through clear instructions setting up the next part of the work, and then leaving

---

206 WYJS/CA MIC2/16 Journal 5th-18th Sept 1763.

207 WYJS/CA MIC2/4, 14th March 1763.

208 Skempton, Ch V, 'Rivers and Canals', Charles Hadfield, London, 1981, p 108.

209 LI 18th June, 1765. Letter to newspaper.

210 Skempton, *John Smeaton FRS*, Ch X 'Professional Practice', Denis Smith, London 1981, p 225.

211 Skempton, *John Smeaton FRS* Ch 1 'John Smeaton', Turner and Skempton, London 1981, p 22.

the day-to-day supervision in the hands of the resident engineer. The commissioners were unconvinced about the need for this last role and when his deputy was removed in 1761 he was not replaced. However, Smeaton never wavered from the structure of project management which he drew up for the Calder Navigation and this was specified again in his document in March 1768 'A Plan or Model for carrying on the Mechanical Parts of the works of the canal from Forth to Clyde'.[212] It remained as the basis of his approach in later projects on which he was superintending or consulting engineer or when he recommended others for such work.

Although he did not describe himself as such at this stage of his career, the role he was deliberately trying to construct was that of the professional engineer, able to command a good salary because of his proven specialist knowledge and excellent planning, drawing and communication skills. Indeed it was the 1760s which saw the emergence of the profession of civil engineering with an increasing number of men providing consultancy services and management of works, working with resident engineers and contractors. Smeaton became known as the first civil engineer, encouraging this hard-won professional approach among others. At the beginning of the 1770s the Society of Civil Engineers was formed. Smeaton attended when he could, and some time after his death the Society became known as the 'Smeatonian Society' in acknowledgement of his contribution to this forum for collaboration and friendship.[213] Smeaton's archive in the Royal Society is evidence of his lifelong desire to record his projects and learning for the benefit of other engineers, many of whom worked with Smeaton or could study his reports, designs and findings.

Amongst the Society's early members were men who had been employed under Smeaton on the Calder, absorbing his approach to engineering projects, and who often worked with him later in their careers. One such was his colleague, Joseph Nickalls, who had returned to London in 1761 and by 1762 was employed to build the Stratford Mill and the waterwheel at London Bridge, both to Smeaton designs. By 1771 he was appointed engineer to the Thames Commission. Another member was William Jessop, whose apprenticeship at Austhorpe drawing up the Calder plans laid the foundations for his own very distinguished career in civil engineering. He acted as resident engineer for Smeaton's plans for fen drainage at Potteric Carr, between 1765 and 1774, and at Hatfield Chase between 1776 and 1783. He first became a consulting engineer in 1773 and gained a reputation in the field of canal and navigation work in the later decades of the century to match that of his old master. John Smith and Co are mentioned in the Journal in October 1763 as working to repair a hole in the skirt of Lupset Dam and the accounts show that the firm was paid in 1761 for Netherton quarry work and in 1763 were employed for unspecified work on several other occasions. This may have been the

212 Skempton, *John Smeaton FRS*, Ch X 'Professional Practice', Denis Smith, London 1981, p 225.
213 Skempton, *John Smeaton FRS* Ch 1 'John Smeaton', Turner and Skempton,London, 1981, p 24.

John Smith who had acted as Chief Engineer on the Don from 1731 to 1766 or his son, also John Smith. In 1767 John Smith junior, trained by his father, became the resident engineer under Jessop for the River Ure Navigation and Ripon Canal, and at a similar time he was the overall engineer on the Linton Locks on the Ouse and Swale and Bedale Beck work, a scheme for which Smeaton gave evidence.[214] He joined the Smeatonian Society in 1772. John Longbottom's name appears in the commissioner's minutes for December 16th 1761 when he submitted a proposal for building a lock and a Mr Longbottom was paid for attending the commissioners on 18th February 1762, although he does not appear to have become a contractor during the years of the Journal. He may be the same John Longbottom (Longbotham) of Halifax who suggested a canal from the Humber to Liverpool in 1766, surveying and acting as engineer for a Yorkshire section of the resulting scheme for a Leeds & Liverpool Canal.

These engineers might often later meet in advisory roles sometimes representing opposing parties to a scheme. However their Calder links gave them a common background which continued in the Society of Engineers.[215] These men, and contractors like Joshua Platts and Luke Holt, who had worked on the Upper Calder, knew each others' strengths and abilities, and furnished much of the initial expertise required in the explosion of navigation and canal construction in the north of England and elsewhere.[216]

During these years the project, an undertaking on a huge scale for the time, faced many external problems arising from conflicting local demands. Mill owners such as Sir Lionel Pilkington at Wakefield or Mr Banks at Dewsbury and Lillands presented the strongest opposition, but there were many others who did not want the navigation to prosper for fear of losses. The increasing speed of infrastructure development at this time meant that new turnpikes impinged on the work in hand, for example in the case of gravel taken from the river. The Halifax to Wakefield Turnpike was nearly complete, and 1758 saw more Turnpike Acts to improve links between Wakefield to Huddersfield and Sheffield. Raw materials such as stone or wood were in demand from a variety of competitors, and dealing with suppliers was fraught with difficulty as in the case of Fretwell. Lateness of delivery or the timing of production often held work up, for example the two month wait for deals from Hull for building sheds in Dewsbury. The weather not only delayed and set back the navigation work but also affected the access roads which were often completely churned up by heavy carts, creating conflict with local people.

However, it is clear from the Journal that it was the River Calder which presented the greatest challenges to Smeaton. Even today, those of us who live in the Calder valley know

214 Hadfield, *Canals of Yorkshire and North East England* Vol 1, Plymouth 1972, pp 103,111.

215 Garth Watson, *The Smeatonians, the Society of Civil Engineers*, London, 1989, pp 3-21.

216 Ed Skempton and Grimes, *A Biographical Dictionary of Civil Engineers in Great Britain and Ireland*, Vol 1 London 2002, p 333-334.

the river's power and volatility, and the ingenuity required in the planning and construction of flood defences. Throughout the years of the Journal to the end of the decade there were many heart-sinking emergencies, reports of damage, delays, and the inevitable rising costs. In February 1763 Smeaton wrote: 'we must never expect to make anything too strong that has to do with the Calder, and must expect many disappointments in so difficult an Undertaking.' By 13th October 1763 following further damage, he again explained that 'this work has been secured as well as it was in my power ... it is one of those accidents that works of this kind are lyable to; more especially with those situated upon such a violent and uncertain river as the Calder.' In a post script to this letter he wrote to Simpson 'though we meet with many repulses I hope time and resolution will conquer at last'.[217] He never gave up, working steadily in the face of further devastation when he was asked to return to repair the defences in 1767, only to find them destroyed before they were completed in 1768. Smeaton's perseverance on this upland river encouraged other promoters and engineers to imagine further waterways to cross the Pennines.

It seems that both Smeaton and his paymasters had underestimated the river at the outset of the work: the difficulties had been far greater, the financial costs and risks much higher, the time required a continuing commitment, and the personal involvement far more demanding than ever envisaged. By 1776 Smeaton's tone suggests great pride in the eventual achievement, recommending that a visiting Russian canal engineer see the Calder Navigation as an example of 'the making of a river navigable according to its own course and this upon one of the most rapid Rivers in the Kingdom, of its Size, and quantity of water that comes down the Valley in times of floods.'[218]

217 WYJS/CA MIC2/4 1758 -1763.

218 Skempton, *John Smeaton FRS, Ch V, 'Rivers and Canals'*, Charles Hadfield, London, 1981, p 124.

NOTES ON TRANSCRIPTION

This transcription has been made from Smeaton's Journal at the National Archives.

In order to make the content as clear as possible whilst retaining the text several decisions about layout, spelling and punctuation have been taken. The Journal has been separated into the four years that it covers, and the regular record-keeping emphasised by providing the dates of each week and designating a number for the week. Punctuation and paragraphing have been altered in places, but only to aid comprehension.

All spellings have been retained as they were written, which has often meant that the same word is spelt in different ways eg 'cut' or 'cutt' from sentence to sentence. The use of capital letters, mostly but not consistently or exclusively for common as well as proper nouns, has been kept. Abbreviations have often been filled out with missing letters in square brackets, although some straightforward omissions or apostrophes have not been altered if the meaning is very clear.

THE
Superintendant's Journal
of the Works of the River Calder
begun in the Spring 1760

51. Title page of the Superintendent's Journal, 1760

THE SUPERINTENDENT'S JOURNAL 1760

Week 1 May 26th - June 1st

Monday 26th

This Day further inspected the Works and found them as follows viz In the Work Yard a Carpenters Workhouse and Storehouse in one Building, a Smith's shop & shed for Minion in one Building, and a Shed for the Sawyers all completed. NB all the above Buildings are made so as to be taken to Pieces and set up at other Places according as the different Situation of the Works shall require.

The Tools, Utensils, Watchet Lime etc from the Edystone Works are arrived. A considerable Quantity of wrought Aisler[219] from different Quarries in the Yard, and a good assortment of Iron: a considerable Quantity of good Oak, Elm, & Fir Timber with some Alder for Piles, and also a large Quantity of Minion almost sufficient for the Summer's Service in the Yard. Several setts of Tryangles with Windlasses for hoisting, & other necessary Utensils ready prepared.
The Horse Water Screw Engine[220] in considerable forwardness and the Carpenter at Work thereupon, the Molds made for the Quoins and hollow Parts of the Lock[221], and Arches for the fore Bays. The Sawyer's preparing Materials for the Punts[222], a Quantity of Stuff for Plank piling sawn out, and some hands at Work in preparing them for use. The Smiths at Work upon Iron Work for the Engine, a considerable quantity of stone from Widow Hartley's quarry[223] delivered in Fall Ing, some scapelled Backing from Robert Hartleys Quarry upon the Steanard[224].

At the Cutt from Thornes to Lupsit all the Hedges Trees etc grabbed up. Mr Platts and Men employed since his arrival in digging upon the middle Part thereof had made some progress. The Ground in the highest Part being opened and sunk in the Rock to its intended Depth within a few Inches; a considerable Quantity of hard bluish Stone got out of the Cutt, and piled upon an Heap for such Uses as it shall appear proper; the Workmen employed in baring the rest of the Ground as far as the Rock Extends, which is near upon to the Hedge that divides Mr Witton's two arable Closes as per Plan.

A Quantity of wrought Aisler from new Miller Dam at Horbury Mill Pasture, and a Sample of Stone got out of the River upon the Shoal just below. The Tail of Horbury Mill Cutt; a Quantity of Brick led and leading into the Lock close for backing.

---

219 ashlar - finely dressed, worked stone.

220 Archimedes screw to pump water, driven by a horse.

221 the external corners of the masonry, and the 'hollow posts' where the lock gates were inserted.

222 a flat-bottomed boat with shallow draft.

223 Ann Hartley's quarry was at Altofts (Ref: WYJS/CA MIC2/4, 11th Jan 1760)

224 Robert Hartley's quarry was at 'out-Woodside'. (Ref: WYJS/CA MIC2/4, 11th Jan 1760)

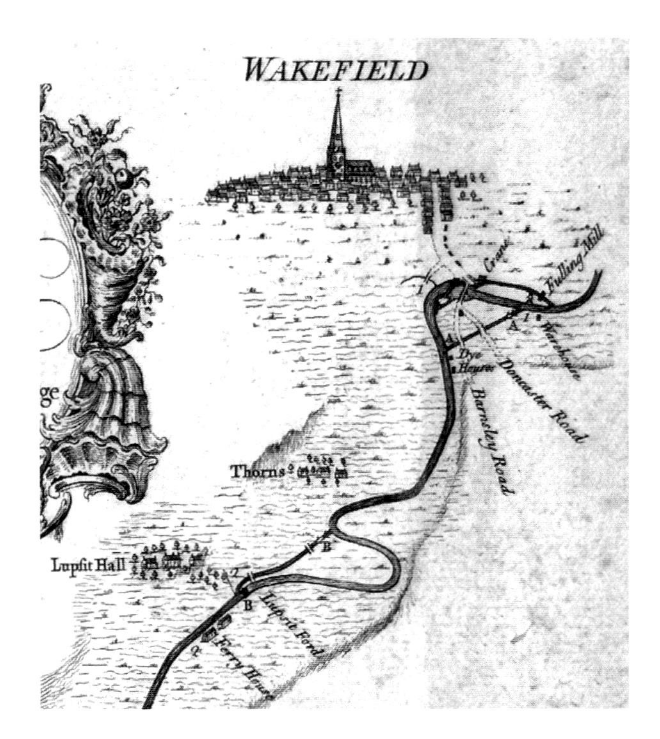

52. *Wakefield and Lupset, from John Smeatons' Plan of 1757/58*

This Day Messrs Smeaton & Nickalls marked out the Ground for the Lockpit at Thorns, and this Day at Noon, the Undertaker[225] begun the diggin thereof; the remainder of the Day employ'd in taking Levels for the last adjustm[en]t of the Depths & Widths of the Cutt at the Different Parts according to the Different Elevations of the Ground and in sounding in the Reach between the Tail of this Cutt and Wakefield, in which occurred some Particulars requiring further Examination.

Tuesday 27th
This day all Hands Employed as yesterday.

Messrs Smeaton & Nickalls in consequence of the Levels taken Yesterday staked out the Cutt for the first time from the Lock as far as the Rocky ground reaches with very little variation from the Plan and former Sequels, so that the diggers may be able to compleat their work as they go on.

This evening Mr Smeaton in company with Dr Jackson[226] took the opportunity of riding to Knottingley Lock[227] to measure the bridge over the same, and at the same time observed the Water running thro' the new Work of the Walls & under the Foundations upon filling and emptying the Lock in a most surprizing manner.

Wednesday 28th
This Morning set forward with Mr Nickalls to attend the Meeting of the Commissioners at Halifax.

Thursday 29th
This Day Mr Smeaton went to Austhorpe in order to forward the Plans & Mr Nickalls returned to Wakefield to forward the Works.[228]

Friday 30th
This Day Mr Smeaton employed at the Plans.

---

225 undertaker - contractor.

226 Dr Cyril Jackson, (1717- 1797) on the committee of Commissioners.

227 Until the construction of the Aire and Calder Navigation, Knottingley had been the highest navigable point of the River Aire.

228 This is the first mention of working plans, which occupied much of Smeaton's time for these years. Some of these still exist in the Royal Society archives, but many have not survived.

Saturday 31st
Ditto

<u>1760 Week 2 June 2nd - June 8th</u>

Monday June 2nd
At Wakefield Hands in the Yard at Work upon the Screw Engine, 2 Hand Pumps, Plank Piles, and a large Punt for transporting Materials upon the Water etc. The diggers had sunk some part of the Lock Pit down to the Level of the Water in Wakefield reach, found there a loose Sand with Water some hands upon the Cutt, the Diggers employed by the Commissioners at Work upon the Rock.

This Day Mr Smeaton in company with Mr Nickalls staked out the upper Part of the Cutt above the Rock to the entrance into the river at Lupsitt Boat[229]; also carefully sounded the Reach from the tail of Thornes Cutt to the Head of Wakefield D[itt]o and in the two Shoalest[230] Places, but best of the Channel found 3 feet in the upper End, 4ft 9in in the general at the lower Cut in one Place 4ft in for a few yards Gravel Bottom. NB the Best of the Channel at the upper Shoal is close Home with the Shoar on the north Side and the lower one with the South Side.

This Day Mr Smeaton having prepared the South Drafts for the Forebay Sternbay Stone Sills for Thornes Lock as also for the Timber Bridges over the Cutt at Thornes & Lupsit. Measures were taken for putting the same in Execution with all possible Expedition Orders having been given some time since to the Cargate[231] Quarry men to lay aside all the largest Blocks of Bluestone[232] they cou'd get for sill Pieces. This Evening Messrs Smeaton & Nickalls and Wilson went to the Quarry to inspect the same, when no one Piece yet raised appeard fit for this Use but being now got down to a Part of large Stuff were in hopes of raising some Blocks large enough in a few Days.

Tuesday June 3rd
This Day Mr Smeaton at Work upon the Plans

Wednesday 4th
Ditto

Thursday 5th
D[itt]o and forwarded to Mr Nickalls fair Copies of the Plans for the forebays Sternbays Sills & Bridges

---

229 possibly the Ferry at Lupset marked on Smeaton's map.
230 shoals - shallow areas of sand or gravel.
231 Carr Gate, 3½ miles north of Wakefield.
232 a good quality sandstone that has a bluish/grey tinge to it.

Fryday 6th
This Day at Wakefield hands in the Yard employed as on Monday... The Lock Pit sunk to the level of the Calder Water in the Wakefield Reach & there stop'd by the Water coming in till the Hand Pumps now preparing are ready; the Hands employ'd upon the Cutt & a small part thereof in Mr Mauds Ground got down to the Depth, Com[missioners] Diggers upon the Rock in the Cutt, part of those with additional hands order'd to the Rocky Shoal at the Tail of Horbury Mill Cutt on Monday next.

This Day & Days past 2 Men employed in shovelling up and carrying away Gravel from the Shoal at Lupsit ford – for the Use of Turnpik[233] so far of Use to us, but if they proceed much further it will be a detrim[ent] in placing the Dam.

This Day Mr Smeaton in Company with Mr Nickalls took the Levels of the two Roads and all other Points of Ground in the Tract of Wakefield Cutt, above the level of the Surface of the Water of Wakefield Reach and found the highest Points of Barnsley Road 3ft 7½in Doncaster Road 6ft 9½in.

Saturday June 7th
This Day at the Plans

<u>**1760 Week 3 June 9th - 15th**</u>

Monday 9th
This Day at the Plans

Tuesday 10th
This Day at Wakefield; the Hands in the Yard employ'd upon the Screw Engine, Punt & Hand Pumps also upon a Platform for framing the Sill Pieces for the Gates upon, and a moveable Cabbin to hold Tools Instruments Papers while the Locks are building; a Mason at Work upon one of the hollow Post Blocks; the Diggers employ'd upon the Cutt as before; a hand scapelling the Stone out of the Rock Part for backing.

This day Messrs Smeaton & Nickalls staked out for the last Time the Lockpit Cutt in Fall Ing in order to ascertain the Quantity to be purchased thereof as exactly as possible.

Wednesday 11th
This Morning being sent for by the Marquis of Rockingham[234] Mr Smeaton went to Wentworth House and took the opportunity of inspecting the Stone needing some reprehension, as also of

233 the turnpike road from Wakefield to Halifax.
234 Charles Watson-Wentworth, 2nd Marquess of Rockingham, (1730 –1782) of Wentworth Woodhouse.

calling upon Mr Fenton of Bank Top[235] who said that under the same consideration as the Rest at Wakefield Lord Strafford[236] gave leave to begin whenever we pleased.

Thurs 12th
This Day return'd to Wakefield, the Work in the Yard as before the Pump being fix'd at the Lockpit at Thornes when one Man was capable of Mastering the Water the Diggers were at work therein. This Week some hands employed upon the Shoal at the Tail of Horbury Mill Cutt.

Fryday 13th
This Day at the Plans

Saturday 14th
Ditto

1760 Week 4 June 16th - Sunday 23rd

Monday June 16th
This Day preparing Plans

Tuesday June 17th
Ditto

Wednesday June 18th
This Morning at Wakefield the hands in the Yard at Work upon the Engine punt & Cabbin. The Platform & the first Stone for the hollow Post completed. The Diggers at work in the Lock Pit being upon a Gravel moderately compact. The bottom about 4 feet below the Surface of the Water in the Contiguous reach the leakage easily pumped out with one of the Hand Pumps, by one Man. Some of the diggers in the Cutt our own Diggers Part at Work getting Rock in the Rocky Part & the Masons scapelling the Stone for backing, and Part employed in dredging the Shoal at the Tail of Horbury Mill Cutt. Ordered the Matter to be led out of the River, & the Side of the Channel to be secured by a foot wharfing of Stakes walled with bindings & backed with rubble Stone. In the afternoon at Halifax being the Day of the Union Club.

Thursday June 19th
This Day at Halifax as also Mr Nichols attending the annual Metting of the Commissioners.

---

235 Richard Fenton of Bank Top, Worsbrough, agent to Lord Strafford.
236 William Wentworth, 2nd Earl of Strafford, (1722 – 1791) of Wentworth Castle.

Friday June 20th
This Day returned to Wakefield where everything going on as per Wednesday; from thence to Austhorpe.

Saturday June 21st
This Day on private Business

<u>1760 Week 5 June 23rd - 30th</u>

Monday 23rd
This Day at Austhorpe forwarding the Plans.

Tuesday June 24th
This Day hard Rain all Day as also all last Night which prevented going to Wakefield.

Wednesday 25th
This Day at Wakefield. Arrived in the Yard and 84 Quarters of Barrow Lime[237] as per Order The Cabbin completed. Workmen in the Yard at the Engine Punt and additional Pump. A Shed for covering the Platform & D[itt]o for the Barrow and other Lime. The piling Rams in Hand. Another hearth fitted up in the Smithy.

In fall Ing observed a Mark which Mr Nicholls reported the Water had been at the preceding Day being about 9 Feet above the Common Level of the Calder.

At Thornes the Flood of the preceding Day had made a Breach at the Corner of the Lock Pit about 3 fathoms wide & at a medium 4 Feet Deep. Mr Nickalls reported that being there when the water overtoped the Bank about 9 o'Clock the preceding Morning being sandy Ground & the Water rising very Quick it cut down the Bank by running over it and did not make its way by carrying it away before it. This is one of those Accidents that must be expected to happen on such sudden Shoals, which are more easily repaired than prevented. The Diggers at work upon the Cutt. The Rocky part filled with Water by the rise of the Water but without apparent Damage. The Mason scapelling Stone at Lupsit ford.

At the Tail of Horbury Cutt no damage as yet Visible.

Thursday June 26th
This Day at Austhorpe forwarding the Plans.

Friday 27th
This Day at Wakefield all Hands employ'd in the Yard as on Wednesday last.

---

237 lime from Barrow on Soar, Leicestershire (Ref: http://www.busca.org.uk/heritage/articles/old-industries-of-barrow/occupations-in-barrow.html)

At Thorns repairing the Breach. Diggers & scapellers as before.

Some pieces of Block Stone came into the Yard this Day from Cargate Quarry for hollow Posts & Sill Pieces, but somewhat inferior in Quality to what was expected from thence for that Purpose. This Evening at Cargate Quarry to inspect the Posts, and at Leemore Quarry[238] of Bluestone belonging to Rob[er]t Hartley where he had bared a Post seeming very proper for hollow Post Blocks.

Saturday June 28th
This Day at Austhorpe forwarding the Plans & at Huddlestone Quarry[239] of most excellent Freestone for the purposes of Sills & Hollow Posts, the Price 10d per Foot cubick at the Quarry & about 3d per Foot Carriage to Ferry Bridge.

Omitted in last Weeks Journal that a sett of hands were employed at Hartley Bank[240] in getting Stone the Quarry appeared very promising another Place in the same Bank was also proposed to be tryed which had a very favourable appearance.

1760 Week 6 June 30th - July 6th

Monday June 30th
This Day at Austhorpe forwarding the Plans

Tuesday July 1st
This Day at Wakefield the Hands in the Yard at Work upon the Engine Platform & Lime Sheds & finishing the Punt which was launched Yesterday, also at work upon the Piles, Piling Engine, & an Engine for groo[v]ing Piles.

At Thorns at work repairing the Breach. The Diggars upon the Cutt. Labourers at Work getting Stone in the Rocky part, and getting out the Water. Masons at Lupsitford scapeling the Stone got out of the Cutt.

This Day Mr Smeaton & an Assistant employed in taking the Levels across Horbury Pasture for the more exact Adjustment of the Works there.
This Day Mr Wilson[241] reported that in all the Places on Hartley Bank lately worked upon, that the Stone died out & became so bad as not to be worth pursuing further.

238 Lee Moor, Stanley, north of Wakefield.
239 This quarry was on the magnesian limestone ridge near Sherburn in Elmet and was of an excellent white quality.
240 south of the river at Horbury.
241 Joshua Wilson, master mason.

This Evening again at Cargate Quarry with Mr Nickalls in order to inspect some Blocks now got up from the lower Beds but which did not seem superiour to the former, the Stone being useable for the purpose of Sills Hollow Posts but not so perfectly compleat as expected.

Wednesday July 2nd
This Day at Austhorpe forwarding the Plans

Thursday July 3rd
This Day at Ferry Bridge being desired by several Gent[leme]n to attend the Meeting there in order to give my opinion upon the Condition thereof.

Fryday July 4th
This Day at Wakefield the Platform & Lime Sheds, Punt & a small Piling Engine completed, two Masons employed upon the Cargate Blocks for sill Pieces; a Second Punt in Hand the grooving Engine[242] going on, at the Lock Pit the Breach repaired & Water pumping out. Diggers Labourers & Scapellers employed as on Tuesday. This Evening drove a Pile in the Lock Pit in order to ascertain the Lengths of the rest

Saturday July 5th
This Day at Austhorpe forwarding the Plans for the Calder Business

<u>1760 Week 7 July 7th – 13th</u>

Monday 7th July
This Day forwarding the Plans for Thornes Cutt and Calder Business.

Tuesday 8th July
This Day at Wakefield being inform'd of a Meeting of the Com[missioner]s of the Agbridge Turnpike attended them on Account of the Navigation Bridge over the Cutt crossing that Road according to my orders at the last meeting, the substance of the Conversation was that the Com[missioner]s of the Turnpike expected that the Road shou'd be detrimented by us the Bridge as little as possible that it shou'd be as Wide as the Road itself not exceeding 9 Yards wide, or otherwise so much wider than Wakefield Bridge as to admit of an Horse causway, besides the breadth of that, that sufficient Battlements or Parapets may be made so that Horses passing may not be frightfull by the passing of Vessels & lowering their Masts that the Bridge itself with its Battlements & the Road itself upon the Bridge for as it may be necessary to pave it or otherwise be kept in repair by the Com[missioner]s of the Navigation that the Road so far as it may be necessary to be raised to render the access to the Bridge easy be done at the Expence of the Com[missioner]s of the Navigation but afterwards repaired by the Com[missioner]s of the Turnpike.

242 probably to apply grooves to the plank piles so that they can be fitted together.

Hands in the Yard at Work upon the second Punt sheetpiling grooving Engine two middle siz'd piling Engines & the Mill for grinding Minion & Pozolana. Masons in the Yard at Work upon the sell Pieces.

At Thorns the Lock Pit being cleared of Water & rubbish by the Navigation's Men the undertakers Diggars at Work getting down the Lock Pit, the springs being encreased particularly from the Sands so as fully to employ 4 Men at the Pumping & likley to encrease on going Deeper determined to put down the Screw Engine. Some of the undertakers Men upon the Cutt & the Navigation Men upon the Rocky Part being now all cleared of Water: Scapellers at Work upon the Stone out of the Cutt at Lupsitford, some Pieces of Bluestone being brought into the Yard by Rob[er]t Hartley for hollow Posts Pieces, inspected the same, which appeared less perfect than expected from the appearance of that Bed in the Qarry.

Mr Nickalls reported that, that Day he had seen Mr Heelis[243] who told him that we might begin to Work upon the Bluestone at Netherton when we would & that the Rent wou'd not exceed two Guineas a Year.

Wednesday July 9th
This Morning Messrs Smeaton, Nickalls & Wilson went to inspect & set out the Blewstone Qarry at Netherton & determined first to get out the Water from an old Working in order to have a better view of the Beds & to set on hands to Morrow Morning.

This Day also inspected the Quarry in Horbury Bank & Hartley Bank & determin'd to try another baring at Hartley Bank, and if not succeed there to begin to bare in Horbury Bank for back[in]g that being all that can be expected there.

This Afternoon Mr Smeaton went to Halifax in order to attend an expected Meeting of the Committee.

Thursday July 10th
This Day at Halifax attending the Adjournment of the Com[missioner]s and reported several matters to the Gent[leme]n who then attended.

This Evening returned to Wakefield & spoke to Mr Nickalls who reported all going on in due Course.

Friday July 11th
This day at Austhorpe forwarding the Plans & Navigation Business.

---

243 Mr Heelis – Mr Fox Lane's steward (Ref: WYJS/CAMIC2/4, letter from Simpson to Mr Wilcock of Halifax, 28th March 1760)

Saturday July 12th
Ditto

<u>1760 Week 8 July 14th – 20th</u>

Monday 14th July
This Day at Austhorpe forwarding the Plans & Calder Business.

Tuesday 15th
This Day at Wakefield, Carpenters in the Yard at Work upon the 2nd Punt, a frame for raising & lowering the Waterscrew, making & fitting large Pullies thereto for turning the same. Masons at Work upon the sell Pieces & hollow Posts. Labourers sifting the Barrowlime & fitting the same for preservation & use.

Carpenters at Thorns Lock setting up the Engine, the Lock Pit being now got down to its general Depth, and 4 Men being unable to keep out the Water Ordered the Pumps to cease till the Engine is got ready. The Undertakers Men at Work upon the Cutt our own Men at Work upon the rocky Part Assisting in setting up the Engine & at the Hartley Bank & Netherton Quarrys.

This Day Mr Smeaton & Nickalls brought down the level afresh; from above Lupsitford to the Tail of the Cutt at Thorns, in order to examine the depth of the Cutt for the finishing thereof.

Finding that Masons were very scarce & not as ready to enter with us as might be expected Agreed by Messrs Smeaton Nickalls & Wilson to give it out that we shou'd keep a Number of Hands During the Winter: and as, so far our contracts for Stone having been served in not greatly to our Satisfaction; and the contractors requiring an exceeding deal of talking to & looking after, agreed also to recommend it to the Committee to engage proper Quarries, & to employ therein the most usefull of our Hands during the Winter in getting Scapelling & hewing Stone for the next Year's Service.

Wednesday 16th
This Day on private Business

Thursday 17th
Ditto

Friday 18th
This day at Wakefield. Carpenters in the Yard at the 2nd Punt Masons at the sells & hollow Posts, Labourers sifting Barrow Lime.

Carpenters at Thorns setting up the Engine, the Screw being got into the Water; Labourers assisting there, getting stone in the Cutt, Hartley Bank & Netherton Quarrys. The Undertakers at Work, Part of the Men in that Part of the Cutt between the Lock & Rocky Ground, Part employed between the rocky Ground & Cutt Head.

This Day Mr Nickalls reported that Joshua Wilson had observed some Men at Work for the Turnpike in the Lane at near Osset street Houses, upon a fine bed of Bluestone. This Day Mr Wilson & Smeaton went to inspect the same & found some Men at Work, about to get some Mile stones & Causway stones for the Turnpike Road from Wakefield to Halifax. Found that the out Brake of the stone is just in the Road side, & that the greatest Part of what has been upon the waste, has already been got from Time to Time for the Neighbours Use and lately for breaking upon the Turnpike.

The Quality of this Stone seems fully equal to that of Netherton, but in a more regular Bed & larger Scantling.[244]

According to inspection & Best Information the Best stone seemed to Run into a close belonging to Mr Herring Coroner at Wakefield: therefore Ordered Mr Nickalls to treat for Quarry leave & agree for any rent not exceeding two Guineas a Year.

Saturday July 19th
This day on Private Business

1760 Week 9 July 21st - 27th

Monday 21th July
This Day at Austhorpe forward the Plans & Calder Business.

Tuesday 22nd
This Day at Wakefield. Hands in the Yard, at the 2nd Punt, Groving Engine, Pozzolana Mill, to Piling Engines & Piles. Masons at the Sells & hollow Posts.

At Thorns the Screw Engine got to work, but not quite completed, Undertakers Diggers upon the Cutt between the Lock & Rocky Ground & also at the Ground from thence to the Cutt head: our own Labourers at Work upon the Rock carrying Stones in the Punts from thence to the Lock for building the Bridge at Thorns.

Masons at Work scapeling Stones at Hartley Bank, & Labourers at Work at Netherton Qarry; this Day went thither to inspect the produce of that Quarry, a quantity of Backing[245] & Setters[246] got but not yet down to the main Blewstone bed.

---

244 the dimensions of thickness, breadth and length.

245 backing – rubble masonry used behind the facing stones.

246 setters – rectangular blocks that were used on the lock pit bottom, and used elsewhere for paving, also known as setts.

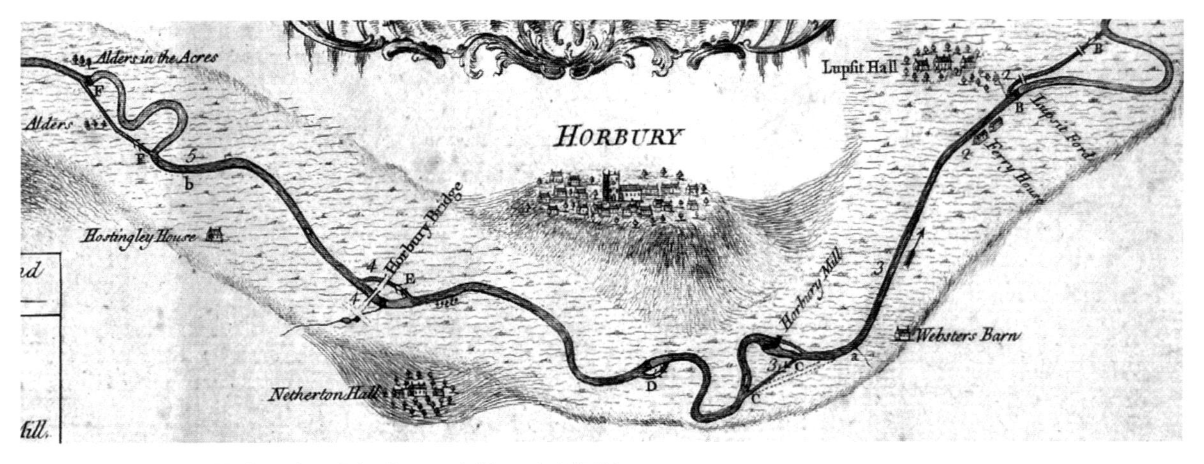

53. Horbury from John Smeaton's Plan of 1757/58

This Day sounded Horbury Dam & Horbury Reach, that is the Reach from Washingstone Ford[247] to Horbury Bridge, in order more exactly to ascertain the Cutts & Works to be erected thereupon.

Wednesday 23rd
This Day on private Business.

Thursday 24th
Ditto

Friday 25th
This Day at Wakefield; Work in the Yard as on Tuesday Last, the Engine Compleated for raising Water and this Day set at Work our own Labourers at the Rocky part of the Cutt, & Quarry as on Tuesday & undertakers at the Cutt Head. This Day we had the favour of Dr Jackson's Company to inspect the Works.

Saturday 26th
This Day Austhorpe upon the Calder Business

1760 Week 10 July 28th - August 3rd

Monday July 28th
This Day at Austhorpe, at the Calder Business.

---

247 Washingstone is marked 'D' on Smeaton's plan.

Tuesday 29th
This Day at Wakefield. The Grooving Engine, & one of the piling Engines compleated. Hands in the yard grooving Sheet Piles, at the Second Punt & Pozolana Mill. Masons at the sells & the hollow Post.

At Thorns the Lock Pit being cleared of Water. This Day the first Gauge Pile[248] of the stern Sheet Pilling was drove, and the matter at about 4 feet below the Lock bottom found very compact. The Engine clears the Water with the greatest ease; that on the last Tryal with Pumps could not be kept clear with 12 Men. The Punt employed in Bringing Lime & Stone from the Yard & Backing Stone from Lupsitford to the Cutt.

At Thorns; Labourers at the Rocky part of the Cutt & Netherton Qarry Undertakers Men at the Cutt head,

This Day Joshua Wilson reported that this Morning the Labourers employed by the Turnpike had again begun to get gravel upon Lupsitford & that pursuant to his orders he had discharged them; that the Master thereupon went to Wakefield to speak to some of the Com[missioner]s of the Turnpike, and at his return proceeded again to Work & gave Joshua Wilson to understand that he should not take any denial, upon going myself to the Ford I found that the Person employed herein is Jas Turner of Dewsbury contractor for repairing the Road from Wakefield to Halifax, & to lay a certain Number of Loads of Calder gravel thereupon by the Year. The Contractors Son being there I informed him that, as we intended to build a Dam upon the Place where they were leading gravel; and as every Load of Gravel that they took away must be replaced by us either with gravel or other Matter of greater expence; and as the Com[missioner]s of the Navigation have an absolute power over the River, I advised him to desist from carrying away the Gravel least he shou'd be obliged to bring it back again & desired him to inform his father, he ans[wered] me that let the Consequence be what it would his father was determin'd to proceed in getting Gravel.

Dr Cookson[249] who happened to be at the Lock, I informed of those proceedings & observed to him how wrong it was that this Man shou'd be encouraged therein it must necessarily breed disputes between the 2 setts of Com[missioner]s to this agreed & said, as he stood by the Lock there was no need of procuring gravel at Lupsitford, when there was such abundance turned out of the Lock Pit; I answered that I cou'd not engage that he shou'd have gravel at the Lock Pit in Lieu of that at Lupsitford, but that as the gravel at the Lock was of no use to us but was in our

---

248 Andy Beecroft suggests that when piling they would have needed to know the length of pile necessary to reach suitable strata at each location in order to cut the working piles to the required length plus a bit. The gauge or gauging piles would be driven in advance of the main piling works at intervals along the structure to find out this information (ie gauge the depth). This would ensure that timber wastage due to excessive pile 'cut offs' would be kept to a minimum.

249 Dr John Cookson of Wakefield (c1700 – 1779) was an active Commissioner for the Wakefield to Halifax Turnpike.

way, I believe he might have as much as he pleased on asking leave and desired the Dr as he was a Com[missione]r that he wou'd mention these Matters to some of the other Gentlemen which he said he wou'd.

Wednesday July 30th
This Day at Austhorpe on private Business.

Thursday July 31st
Ditto

Friday Aug 1st
This Day at Wakefield hands in the Yard, in the Cutt, Quarries & Punt employ'd as on Tuesday last

At the Lock proceeding with two Engines at the piling. Part of the Undertakers Men at Work in that Part of the Cutt which passes through the Workyard in order to make a Convenience of Watering the Timber.

This Day Messrs Smeaton & Nickalls staked out the Ground for the last Time, for the Undertakers Men to begin upon the Cutt & Lock Pitt in Horbury Dam Mill Pasture[250]. Joshua Wilson reported that the Turnpike Men had been at Work upon Lupsitt shoal ever since Tuesday.

Saturday 2nd
This Day at Austhorpe at Work upon the Designs for the Lock at Thorns.

1760 Week 11 August 4th - 9th

Monday Aug 3rd (the dates for this week were wrongly recorded in the Journal)
This Day at Work upon the Plans & Calder Business.

Tuesday 4th
This Day at Wakefield hands in the Yard grooving Sheet Piling, at the Second Punt & Pozzolana Mill, the 3rd Piling Engine completed & at Work, the Masons on the yard at Work upon the Sells & hollow Posts

250 this is later referred to simply as 'pasture lock' and by 1806 is called 'Broad Cut Lock' (Ref: WYJS/WA C299/4/5/2 Plan of the Calder & Hebble Navigation taken in the year 1806)

At Thorns the Piling going on with 3 Engines the Point Pile[251] for the upper Gate Plank Piling drove and the String Pieces fixed thereto The Labourers employed at the Piling in the Punt bringing Stone & Lime from the Yard to Thorns. At the Rocky Part of the Cutt underpinning the Sides of the Cutt, where the Matter is loose, at Netherton Quarry &c
The Undertakers Men at Wakefield Lupsit & Horbury Cutts, this Evening left orders with Messrs Nickalls & Wilson for proceeding till this Day sennight[252] & set forward upon a Journey.

NB The Turnpike Men has not fetched any Gravel since Fryday.

Wednesday 5th, Thursday 6th, Friday 7th, Saturday 8th
Upon my Journey and at Louth in Lincolnshire[253]

1760 Week 12 August 11th - 16th

Monday 11th
This Day returned from Louth.

Tuesday 12th
This Day at Wakefield; hands in the Yard employed upon the Second Punt at the Pozzolana Mill at the Stamper Mill for beating Morter and grooving Plank Piling, Mason at the Sells and Hollow Posts, the Arched Sell Pieces for the lower Gate of Thorns Lock got together upon the Platform,

At Thorns the bearing Piles all completed four Bays in 5 of the Stern Sheeting drove, and the last begun, this Day the String Pieces for the Plank Piling being fixed the Piling was begun, and the first of them caucked which succeeded to expectation, the Sells Pitt for the lower Gates was a sinking the Point Pile drove, 2 of the string Pieces fixed and the Plank Piling begun.

The Labourers at the Piling, in the Punt, at the Sell Pitt, the Rocky Part of the Cutt, and the Quarry's the Undertakers men at the Cutt in the yard, at Lupsit but principally at Horbury Mill Cutt.

This day a fresh[254] in the River, but not so high as to disturb the Works.

---

251 With reference to the drawing of Thornes Lock (RS JS/6/25) the piling is shown in a V shape at both the top and bottom gates. Andy Beecroft suggests that this was installed at depth before the masonry work was placed and appears to show some form of interlocked planking as opposed to the feathered piles, and that in order to fix the point of the V a pile would be required to brace / tie the plank piling into.

252 sennight - seven nights, a week.

253 John Smeaton was consulted over plans for the Louth Navigation by John Grundy the engineer; an Act of Parliament was passed in 1763. (Ref: https://en.wikipedia.org/wiki/Louth_Navigation)

254 fresh (freshet) - a sudden rise in the level of the river due to heavy rain or snow melt.

Wednesday Aug 13th
This Day myself and Servants at the Plans for Horbury Mill Lock and Calder Business

Thursday 14th
Ditto

Friday 15th
This day at Wakefield Pozzolana Mill compleated and the Pozzolana arrived. The other hands employed in the Yards as on Tuesday last,

At the driving the Plank Piling at the stern Gate Sells, this Work had been somewhat retarded by the great inequality of the Matter which had carried the Plank Piles out of upright and required a Particular Provision to reduce to rights. This Day the first stone was Laid by the Mason beginning to Pitch the rough Setters between the Heads of the Bearing Piles[255].

The Work in the Cutt Qarry etc as on Tuesday.

This Day we were favour'd with the Presence of Dr Jackson & Mr Prescott

Saturday 16th
This Day at Austhorpe at the Plan's

1760 Week 13 August 18th - 23rd

Monday 18th
This Day at Austhorpe at the Plans & Calder Business.

Tuesday 19th
This Day at Wakefield hands in the Yard at the second Punt, Stamper Mill and grooving Plank Piling; Masons at the Sells & hollow Posts.

This Day Messrs Smeaton & Nickalls laid down the Lines upon the Sell Pieces (being put together upon the Platform) for forming the threshelds & Centers for the Stern Gates at Thorns Lock.

The Hands at Thorns upon the Stern Gate, Plank Piling which was considerably advanced, being drove into very hard matter; and fixing the Pozzolana Mill. The Masons at Work fixing the rough

255 a series of bearing piles were placed at the bottom of the lock pit to provide a solid base.

Setters for the foundation to stand upon, Labourers at Work at the Piling engines, in the Punt bringing Materials from Wakefield & Lupsit & at Netherton Quarry, a few Hands employed at Lupsit scapelling Stone got out of the rocky Part of the Cutt, which is almost completed, but the Labourers being diminished on account of Harvest, & the rest wanted to the Piling & other Works, from a very few that have of late been employed there are now diminished to none. The undertaker's Men a few at Wakefield Cutt, a few more finishing the Cutt head at Lupsit, & the principal Part upon Horbury Mill Cutt

This Day Messrs Smeaton & Nickals levell'd the River from Washingtone ford to Horbury Bridge & also more particularly what might relate to Mr Lane's Mill[256] there from whence it appears that this Mill will suffer no real Injury from carrying on the Works in their Reach nearly as laid down in the original general Plans.

This Day also inspected Netherton Quarry where there is a promising appearance a good deal of Stone being got from the upper Beds fit for Caping; Ashler Setters etc but none yet of sufficient Bulk for Sells & hollow Posts.

Wednesday Aug 20th
This day at Austhorpe self & serv[an]ts[257] at the Horbury Plans

Thursday 21st
This day absent on Private Business. Serv[an]t at the Plans.

Friday Aug 22nd
This Day at Wakefield.

This Morning hard Rain which indicated a fresh in the River Calder, but in going over Wakefield outwood observed the Water not mended at Lake Dam, whence concluded the Rain had not been so severe in the West: however on going down to Wakefield Bridge, after stopping a little while in the Town found the Water 7 or 8 feet higher than usual, in the Tail of Wakefield Dam Mr Nickalls being just return'd from the Lock where he fortuneately was when the fresh begun to come, he gave account that from the first appearance the Water rose four feet perpendicular in 20 Minutes, at the Tail of Thornes Lock, and that he had found a necessity of filling the Lock Pit, by means of a trunk[258] which had been prepared & laid into the Dam for the Joint Purpose of discharging the Water from the Engine or filling the Pit by topping from the Calder on the appearance of Floods; which preventing the Water from filling the Pit by Topping the Dam in consequence there seems not the least of appearance of Damage to the Works.

---

256 George Fox Lane of Bramham, (c1697-1773).

257 Smeaton is likely to be referring to William Jessop (1745-1814), his apprentice.

258 trunk - drain.

This Afternoon Messrs Smeaton & Nickalls went up from the Lock to Lupsit & Horbury Cutt to observe in what manner the Water might effect the new Banks, but the Water not having been as High has the Grass there & now much abated, could conclude nothing yet saw no reason for apprehension.

The Lock Hands employed in the Yard at the Works as on Tuesday last; the Hands were not beat out of the Cutt either at Horbury or Wakefield.

Saturday August 23rd
This Day at Austhorpe self & serv[an]t employ'd at the Plans as before.

1760 Week 14 August 25th - 30th

Monday 25th
This Day at Austhorpe self & Serv[an]t at the Plans and Calder Business

Tuesday 26th
Ditto

Wednesday 27th
This Day at Wakefield the 2nd Punt launched & fitting up her Masts & rigging; hands employed in grooving Plank Piling and at the Stampers, the Masons at the Sells & hollow Posts, the upper Gate sell Pieces having been together upon the Platform, marked out & the Thresholds working.

At Thorns the Water out of the Lock Pit since Monday. The Hands proceeding with the Plank Piling under the Gates, that for the stern Gates nearly compleated, & proceeding with the forebay; The Pozzolana Mill made ready for use; Labourers at the Piling in the 1st Punt & clearing a foundation for a Bridge over the Cutt near the Lock for Communication with the Thorns Lands, & at Netherton Quarry.

This afternoon at Halifax at the Union Club.

Thursday 28th
This Day at Halifax attending the meeting of the Com[missioner]s

Friday 29th
This Day at Wakefield the Works at the Yard as on Wednesday the Sheet Piling for Lupsit Dam being in hand.

At Thorns the Stern Gate Plank Piling being compleated the Masons at Work rough setting the foundation for the Stern Bay; one of the Wings of the Plank Piling of the forebay compleated & the last going on Labourers employ'd as on Wednesday.

Saturday 30th
This Day at Wakefield attended Sir George Savile[259] who came to inspect the Works which were going on as Yesterday.

1760 Week 15 September 1st - 6th

Monday Sept 1st
This Day at Wakefield to attend a meeting of Com[missioner]s of the Weeland Turnpike[260] on Account of the Navigation Bridge where the following minute was made of the agreement.

'Wheeland Road

Length of the Navigation Bridge to the end of the Battlement	77 feet
Width of D[itt]o. out & out	24
Length of the Road in the middle of the Bridge to be paved & the Horse causeway to be continu'd	34

It is proposed by the Com[missioner]s of the Calder Navigation to make the Bridge & the whole of the Alteration of the Road at their own Expence and to keep in Repair the Stone Work & the Battlements thereof together with the Pavements extending 34 Feet Length in the middle; and so much of the Horse causeway as is included in the same Length of 34 Feet pavement the Commissioners of the Turnpike to keep the gravel at each end & Horse causeway so far as the Gravel extends in Repair up to the Pavement'

Tuesday Sept 2nd
This Day at Wakefield Hands in the Yard grooving Sheet Piling for Lupsit Dam & at the Timber Work for Thorns Bridge Masons at the Sells & hollow Posts.

The Plank Pilling at Thorns Lock being compleated & Rough setters under the Stern Bay compleated & drove down,

This Day prepared a Qantity of Cimment, & began to set the first Course of the Counter Arch, of the Stern Bay, and gave Joshua Wilson Instructions In writing for making & compounding the Ciment proper for the different Uses.

259 Sir George Savile (1726-1784) of Thornhill, MP for Yorkshire 1759-1783, a Commissioner.
260 road from Weeland near Snaith to Wakefield - the road that comes from Pontefract into Wakefield.

Wednesday Sept 3rd
This Morning at the Lock to see the Work in a way to go on properly & this Afternoon went to Wentworth House[261].

Thursday 4th
This Day at Wentworth House.

Friday 5th
This Day returned to Wakefield by way of Thorns the Mason is at Work upon the Masonry of the Stern Bay and at the Rough setters the forebay Hands in the yard employed as on Tuesday

Saturday 6th
This Day at Austhorpe at the Plans

N.B. The Length of 77 Feet at according to the Plan becomes necessary on account of the obliquity of the Canal to the Road, for the Terminations of the Battlement exceeds the side of the Canal only 8 feet.

1760 Week 16 September 8th - 14th

Monday 8
This Day at Austhorpe, self & Servant at the Plans & Calder Business.

Tuesday 9
This Day at Wakefield, hands in the Yard making an additional Trunk for letting the Water into the Lock Pits, when it rises to above a certain Height preparing Piling for the Dam; making a small Thames Punt; & an Hand Pump for Horbury Cutts which proves very wet below the Level of the Calder; Masons in the yard preparing Stone for the chamfered Curve[262]; This Day set the first being the joint Stone[263] for the Arched Sell for the Stern Gates of Thornes Lock, & begun to set the Arch Stones in the forebay some Hands employ'd in ramming down, & compleating the Rough setters in the foundation, & some others rough setting in the foundation for Thorns Bridge.

Wednesday 10
This Day at Austhorpe self & Serv[an]t at the Draw[in]gs.

Thursday 11th

---

261 Wentworth Woodhouse, home of Charles Watson-Wentworth, 2nd Marquess of Rockingham.
262 the 45° angle masonry where the side walls of the lock meet the lock bottom.
263 joint stone - a seating stone off which to build the arch.

Ditto

Friday 12th
This Day at Wakefield hands in the Yard employ'd as on Tuesday last; this Day at [sic] put in the joint Stones, and Grouted up the joints of the sell pieces which being now together & the Stern Bay compleated; begun to Ram up the vacancies round about it, with Gravel & Corn mold Earth[264] the rest of the hands in the Lock and at the Bridge as on Tuesday last, this Day Messrs Royds[265] fav[our]'d the Works with their inspection.

Saturday 13
This Day at Austhorpe, employ'd upon the Design.

1760 Week 17 September 15th - 21st

Monday Sept 15th
This Day at Austhorpe employ'd upon the Designs

Tuesday Sept 16th
This Day at Wakefield. Hands in the yard at the Thames Punt at the Trunk for the Water making Tackle Blocks. Prepar[in]g Timber Work for Thorns Bridge. Piling for Lupsit Dam.

Hands at the Lock compleating the rough Setting, and going in with the Arches of the Forebay. Some Hands at Work upon Thorns Bridge, and pumping the Water out of the Rocky Part of the Cutt in Order to compleat it.

This Day view'd Horbury Cutt which turns out very Loose Gravel and Sheet Sand, and very Leaky. View'd Hartley Bank Quarry, Appearances there not very favourable, but some good Backing will be got; also view'd Netherton Quarry, which is now down to the main Bed, and very promising for Blocks for Sill Pieces and hollow Posts.

This Day Messrs Stansfeld and Cooke[266] favor'd the Works with their Inspection.

Wednesday 17th
This Day at Austhorpe self and Servant at the Designs.

Thursday 18th
Ditto

---

264 corn-mold - probably what is now termed 'loam'.

265 Jeremiah and John Royds were commissioners.

266 David Stansfeld and Benjamin Cooke, both commissioners.

Friday 19th
Ditto

Saturday 20th
Ditto

Sunday 21st
Ditto

1760 Weeks 18 - 22 September 22nd - October 26th

Monday 22nd
This Day at Wakefield Hands in the Yard at the Thorns Punt, the Tackle Blocks, the Timber Work for Thorns Bridge, the Piling for Lupsit Dam, and making a Syphen for draw[in]g the Water over the Stern Gate Sill.

In the Lock the Forebay being completed to the Chamfer'd Course, and this Day the Sett[in]g being compleated. This Even[in]g fetched the two first Stones in the Side Walls. The Ramming round the Stern Gate being Render'd somewhat patchy by the Action of the Spring below, order'd a Sheet of rabbeted Piling on each Side, the Lock below the Stern Gate Sell to extend about 5 feet, and some bearing Piles among the Setting.

Thorns Bridge advancing. This Day delivered to Mr Nickalls the Remaind[e]r of the Plans for Lupset Dam, Thorns and Horbury Locks, which compleat the Set of Designs for Thorns Lock and 2 Carriage Bridges, Lupset Dam, Horbury Lock and the Road Bridges over Wakefield Cutt, and in Order that they may be on Occasion referr'd to by the Com[missione]rs and Gent[leme]n of the Com[mitte]e the following is a Catalogue of the Designs deliverd to Mr Nickalls since my arrival into Yorkshire.

1st	The Plan of the Foundation of Thorns Lock.
2	A Plan of the upper Works of D[itt]o
3	A Transverse section of D[itt]o
4	A Longitudinal Section of D[itt]o
5	A Plan of & Section of the Sell Arches and Counter Arches to a Large Scale.
7th	A Plan and Section of the 2 Carriage Bridges over the Cutt from Thorns to Lupsit
8th	A Plan and Elevation for erecting a Pair of Sheers for Loading and Unloading the Punts at the Work Yard.
9	A Plan and Section of the Cast Iron Boxes serving as Gudgeons[267] for the foot of the Lock Gates
10th	A Plan and Elevation of 2 Road Bridges over Wakefield Cutt.

267 gudgeon - the tubular part of a hinge into which the pin fits to unite the joint.

11th A Plan for the caping of the Gate Piers showing the manner of fixing the Collars[268] for the Top of the Gates.
12th Two Elevations for the Gates of Thorns Lock.
13th A Plan for the Foundation of Horbury Lock
14th A Plan of the upper Gates of D[itt]o
15th A Transverse Section of D[itt]o
16 A Longtitudinal Section of D[itt]o
17th A General Plan for Lupsit Dam
18 A plan of the Draw Gates & one Bay of the Piling and Timber work for D[itt]o to a Larger Scale.
19 A Longtitudinal Section of the Draw Gates and Bay of Timber Work and a transverse Section of the Sluice form'd by the Draw Gates
20 A transverse Section of the Dam Stones piling and Timber Work.

Tuesday Sept 23rd

Having now prepar'd & explaind the necessary Plans, seen a Specimen of every part of the Work perform'd, the whole Laid out, the Machinery in action, and given the necessary Directions for proceed[in]g therewith to Messrs Nickalls & Wilson, this forenoon return'd from Wakefield to Austhorpe, and this afternoon preparing to set out for Scotland.

Week 19 29th - 5th Oct, Week 20 6th-12th Oct, Week 21 13th -19th Oct, Week 22 20th -26th Oct. *During these weeks Smeaton was in Scotland visiting Dumfries and Glasgow.*

1760 Week 23 October 27th - November 2nd

Saturday November 1st

On Saturday the 1st inst arriv'd at Austhorpe from my North journey, and on Sunday Mr Nickalls came over who gave me an Account of the Works.

1760 Week 24 - 25 November 3rd - 9th

Wednesday 5th, Thursday 6th

At the Works the Hands in the yard at Work upon a second Water Screw, and making Tackle Blocks; the Stuff preparing for the Lock Gates. The Timber Work for Thorns Bridge compleated and put together in the Yard, the small Thorns Punt, and an additional piling Engine compleated.

The Diggers at Work upon Wakefield Cutt chiefly between Fall Ing and the Doncaster Road; preparation making for executing the Bridge upon that Road.

268 the upper part of the lock gate is secured to the side-walls by an iron collar, within which the post turns.

Some Hands baring a Quarry in the Side of the Barnsley Road about 200 yards from the Cutt, in Order to get Backing for the 2 Bridges, and afterwards for the Lock in Fall Ing, to which it will lay very gain[269] when the Lock Cutt is completed.

At the Lock the Lock Pit dry which was filled by a Flood the Day before. The Walls rais'd about one Course above the 3rd Sett off that is ab[ou]t 3½ from the Top; but the Fore bay and Counter Arch forward terminated at the 2nd Course above the Chamfer, and the Foundation of the Forewings not Laid.

At Thorns Bridge one of the Side Walls rais'd to its Height, the other wanting about 3 Feet. Determined not to carry up the Lock Wall any higher till the Weather shoud turn out more dry, Least a Frost succeeding shou'd do the Damage, but to proceed with setting the Lock bottom, and to employ the rest of the Hands upon the Bridges, and in hewing and scapelling Stones at the Quarry

At Horbury Mill Pasture a great Part of the Tail Cutt compleated but which being in general a very Loose Sand the Floods have occasion'd it to cove in several Places. The Lock Pit down within 3 Feet in several Places but the whole fill'd with Water, the Men having been quite beat off by the late Floods, determined to carry up a Trench where the Cutt is not compleated in Order to take off the Water as deep as possible and to put on the Screw as soon as compleated in order to raise the Water from the Bottom being a very spungy Lock Pit.

The Quarry at Hartley Bank quite worked out; at Netherton, or as Workmen chuse to call it the Coxley Quarry a Large Quantity of excellent Stone rais'd thereat, and a still better appearance for Sells and Hollow Posts.

A new Quarry open'd at a place sometime ago pitch'd upon in Horbury Bank called Adingforth Pasture which is Likely to turn out better than at first expected being Likely to supply a Large Quantity of Backing upon very moderate Terms being near the River. At the Quarry met Mr Carr[270] and two others of Horbury who came to make Proposals. They ask'd 9d and afterwards 6d per yard superficial of Ground including all its Contents as deep as we shoud be inclin'd to go to which we bid 2d that is at the Rate of Upwards of forty Pounds per Ton to which they did not think proper to consent.

Friday Nov 7th At Austhorpe employ'd upon the Calder Business

8th Set forward for Staffordshire and return'd on Thursday the 20th

---

269 gain - close to.

270 Robert Carr of Horbury.

Week 25 10th Nov - 16th Nov. *During this period John Smeaton met James Brindley in connection with proposals for what would be the Trent and Mersey Canal.*[271]

1760 Week 26 November 17th - 23rd

Saturday 22nd
At Wakefield yesterday was the Highest Flood that had been this Season having made its Way thro' the Work Yard, but don't find that it has done any Damage worth remarking further than filling the Works.

Part of the Masons at Work in the Yard upon the Sell pieces and hollow Posts, the Rest being distributed at the 3 Quarries at Barnsley Road Side, Adingforth and Netherton Quarries.

The Carpenters at Work upon the new Water Screw and at the Tunnel for Letting of the Water off a ditch intersected by the Cutt and Fall Ing so as to pass off underneath the Cutt in Order to prevent the adjacent Grounds and Gardens from being otherwise affected by the Floods than they have been heretofore.
The Diggers have not been able to proceed with the Netherton Cutt since my Last, and at present are beat out at Wakefield by the present Flood.

A considerable progress made with the Lock Pit at Fall Ing, and a Large Trunk set for carrying of the Drainage from the Pumps, and for Letting in of Water in Time of Flood.

Some hands employ'd in making a temporary Road during the Building of the Weeland Road Bridge.

1760 Week 27 November 24rd - 30th

Tuesday 25th
This Day at Wakefield.

The Masons and Carpenters at Work in the Yard and at the Quarries as on Saturday last.
The Diggers at Work upon the Cutt and Lock Pit in Fall Ing, some Hands employ'd in making a Drain from the Calder for conveying the Water from the Tunnel for Drainage before mention'd. The Pumping now grow[in]g very heavy, and the Lock Pit being a very open Gravel

271 Smeaton had previously met Brindley in 1758 about a canal upwards from Wilden Ferry, and met with him again in this period with Hugh Henshall, approving their suggestions with some amendments. Smeaton met with Brindley again in 1762 over the Duke of Bridgewater's proposed extension of his canal to Runcorn. (Ref: Rivers and Canals, Charles Hadfield, Ed. Skempton, London, 1981, p 110)

and consequently very leaky, determin'd to bring down the Screw Engine from Thorns, for as the Work under the level of the Calder there is for much the greatest part dispatch'd, and the Masonry far advanc'd, the principal Engine will be wanted next in Course at Wakefield,

The temporary Road now compleated, and several Hands employ'd back at Barnsley Road Quarry, not only in scapelling Stone for the same but in making Aisler for all such Parts of this and the other Bridge as shall be under Water.

Wednesday Nov 26
This Morning inspected the Grounds thro' which Horbury Bridge Cutt is intended to pass, and from thence proceeded to Halifax.

Thursday 27th
This Day at Halifax

Friday 28th
D[itt]o and returning to Austhorpe

1760 Week 28 December 1st – 7th
(the dates for this week were wrongly recorded in the Journal)

Thursday Dec 5th, Friday 6th
At Wakefield Yesterday again an high Flood, that made its Way thro' the Work Yard, being nearly of the same Pitch as that which happend the 21st Ult. This Flood filld the Lock Pit at Wakefield and the Works with Water, but without any material Damage.

The Masons at Work in the Yard converting into Sells, hollow Posts & Coinstones all the remain[in]g rough Blocks that had been brought from Car Gate and Lee moore Quarries, and in getting the Sell pieces upon the Platform, in order that the same may be remov'd to Netherton Quarry.

The Carpenters in the Yard employ'd in making Moulds for the Stone Work, and in fitting up an additional Broad wheel'd Cart having found our Former one so extremely useful and ready as to render it very desirable to have an additional one, that one may stand to fill while the other is Leading.

Thursday Evening the fixing of the Screw Engine at Wakefield Lock was completed, and the Calder being fallen below the Level of the Trunk, the next Morning set to Work w[i]th the new Screw.

The Workmen having found it too troublesome to disengage the Screw Frame from the Lock Pit at Thornes were obliged to leave it, but having got the Screw on Board the Punt with some weightier Matters just before the Flood came on, the Punt was by some Means broke loose and sunk, and the Screw floating down with the Punt Stream was laid hold of in Fall Ing and secured.

As to the Punt there is no Fear of recovering it as soon as Occasion permits, this is the most considerable Accident that has happen'd to us during the Flood.

Having lookd over all the Works both at Thornes Lupsit and Horbury Mill have found much less Damage than coud be expected being nearly in the same Situation as I found them on my Return from my North Journey as I mention'd in my Journal of the 5th and 6th ult.

All our Masons except those in the Yard employ'd as already mention'd, are at Work at the 3 Quarries in hewing and scapelling of Stone of which a large Quantity is rais'd for all purposes at Netherton, where being furnished with proper Work Shades[272], Bellows and Anvill for sharpening their Tools the Work is going on very briskly. The Masons being now at Work upon Hollow Post Stones and Dam Stones. Also at Adingforth Quarry tho[ugh] open'd with a View To get Dam Stones, and Backing for Horbury Bridge Lock and Dam, yet is likely to turn out a Quarry for good Aisler upon w[hi]ch some good Samples have been work'd by the Masons, and are now proceeding therewith, a Quantity of Dam Stones and Backing being already let to task, and the rest being intended to be put in the same Way as soon as the Value can be ascertaind by proper Tryals.

The Work at Barnsley Road Quarry is also let in the same Manner upon advantageous Terms.

1760 Week 29 December 8th -14th
(the dates for this week were wrongly recorded in the Journal)

Thursday 12th, Friday 13th
These Days employ'd in setting out the Lock Pits and Cutts of Washingstone and Horbury Bridge conformable to the Order of the Com[missione]rs at their last Meeting[273], and considering that the Matter of the locks of the Calder turns out on many Places extremely loose being Nothing but the Wreck of Former Courses of the River. I thought it proper to set out these two Last Cutts with somewhat greater Slope than the former, which was that w[hi]ch has been generally us'd by Engineers on like Occasions.

The Diggers at Work on Part of Wakefield Cut and Doncaster Road Bridge Pit. Mr Nickalls reported that since my last the Discharge of Water has been so great into the Lock Pit that the smaller Screw which had been put down was not able to overcome the Water; and that therefore he was preparing to put down the larger Screw that has been employ'd at Thorns, he also reported that he had Let Bargains to the Masons to work Stone at the foll[owin]g Prices

272 work shades - probably protection from bad weather.

273 At a meeting of Commissioners Nov 27th 1760 it was *'order'd that the Cuts at Washingstone Ford and Horbury Bridge be set out wide enough for two boats to pass'*
(WYJS/CA MIC2/2 Commissioners' Minutes)

Barnsley Road Quarry

d
Aisler 0. 1½ p[er] F
Scapelld Back at 1. 6 p[er] Yard
Wall Stones at 2. 6 p[er]Y[ar]d

Addingforth Quarry

d
Aisler 0..2 p[er] F
Damstones 0..2 d[itt]o
Scapd Back 1..6 p[er] y[ar]d
Wall Stones 2. 0 d[itt]o

Netherton Quarry

d
Aisler 0..3 p[er] F
Damstones 0..3 d[itt]o
Hollow post
Stones 0..6 d[itto]
Wall Stones 2..0 p[er] y[ard]
It is to be noted that the Principal Purpose of working Netherton Quarry is for hollow Posts and Sell Pieces, but as in getting these a large [sic] a large Quantity of Stone got for Aisler, Damstones, and Backing is necessarily raisd it is better to work them tho' at an advanced price, than to throw them away, the advance'd price being necessary on Acc[oun]t of the Hardness of all the Stone of this Quarry.

The Afternoon of this Latter Day went to Halifax being sent for there on Acc[oun]t of the Water.

1760 Week 30 December 15th - 21st
(the dates for this week were wrongly recorded in the Journal)

Thursday 19th, Friday 20th
These Days at the Work…Mr Nickalls reported that on Tuesday Night a Flood had arose as high as any of the Former w[hi]ch had been attended with the usual Consequences: viz that of filling all the Works with Water without any further Damage as yet apparent; That in Consequence since the Weather was so very unsettled, and Christmas coming on; Mr Platts had desired Leave to take off his Men till after Xmas or till the Weather shou'd appear more favourable; which was granted, and also Leave for such of the Carpenters and Masons as choose it, to leave the Work

for the Holidays, on Acc[oun]t of the Shortness of the Days, and Unaptness of the Season for all Kinds of Works.

at p[re]sent some Hands fixing down the great Screw at Wakefield Lock and some Masons at Work finishing such Sells and hollow post Stones P[ie]ces as remain in the Yard The Last Arch in the Yard being now upon the platform.

Friday Morning Messrs Smeaton and Nickalls proceeded to take a complete View of the Grounds, from the Point marked H in the general Plan of the Calder opposite Dewsbury Church to the Tail of the Figure of 3 Lock Cutt mark'd E in order to examine into the Practicability of drawing our General Plan Canal from the Point H to the Point E without going into the River between the said Points by which Means the Dam at the Head of the Cutt E will be avoided and consequently all Pretence for Alteration concerning the Effect thereof upon Dewsbury New Mill: the crossing the River at Heaton Town, and the interfering with the Dewsbury new Mill Cutt, as well as some Lesser Impediments will be avoided. And we are jointly of Opinion that a Canal may be carried along the Skirt of the Mill on the South Side of the River and by means of 3 Locks to drop into the Tail of the Cutt E by which Means the Cutt will in the gen[era]l pass thro' Grounds of less Value than those of the Water Side Meadows, and will pass to the South of Thornhill Lodge.[274]

An Estimate of the Expense of doing the Work each Way, I shall furnish the Committee with as soon as I have Leisure to do consider the Same. NB All the Land through which this as well as the former Cutts at F & H will pass is at present Lady Savile Sir George's Mother married to Capt Wallace and at her Death comes to Sir George.

Also inspected the Quarry under Thornhill just opposite to the Alders in the Acres out of which Blocks of any Size may be cutt, but the Stone being of a much harder Nature than that of Addingforth it is apprehended will be more expensive working. However as this Quarry lays within a Furlong of the Alders in the Acres (and in Case the Cutt now mention'd shou'd take Effect) it will pass under the very Brow where the Quarry is. We think the nearness of its Situation will more than compensate for the Hardness of the Stone, and therefore determin'd to make a Tryal thereof shortly.

In my Return inspected Netherton Quarry, where they were cutting out some of the finest Blocks of Blue Stone I ever saw. A Quantity of excellent hollow Posts and Pieces work'd as well as Aisler and Damstones, but had the Mortification to find the same Spirit reigns thro'out, viz how to avoid a strict Performance of their Bargains for tho' they had themselves worked Samples of each kind, yet the Moment they went upon Task Work the Work was done in an inferior Manner & I therefore told the People and order'd Mr Nickalls to pursue the same that Let the Bargain be what it will if they fall short of the Sample to pay them in Proportion.

274 The Long Cut was an alteration to Smeaton's original plan, changed due to the problems with floods.

1760 Week 31 December 22nd - 28th

Monday 22nd

This Day at Wakefield where Nothing material occur'd concerning the Works themselves more than has been already mention'd being now nearly at a Stand except the Stone Workers at Task Work. Therefore leaving Left the necessary Orders with Mr Nickalls took leave of the Works for the present Season and herewith conclude this Years Journal wishing all Health Prosperity and the Compliments of the Season to my honour'd Masters and am their most humble Serv[an]t

J Smeaton

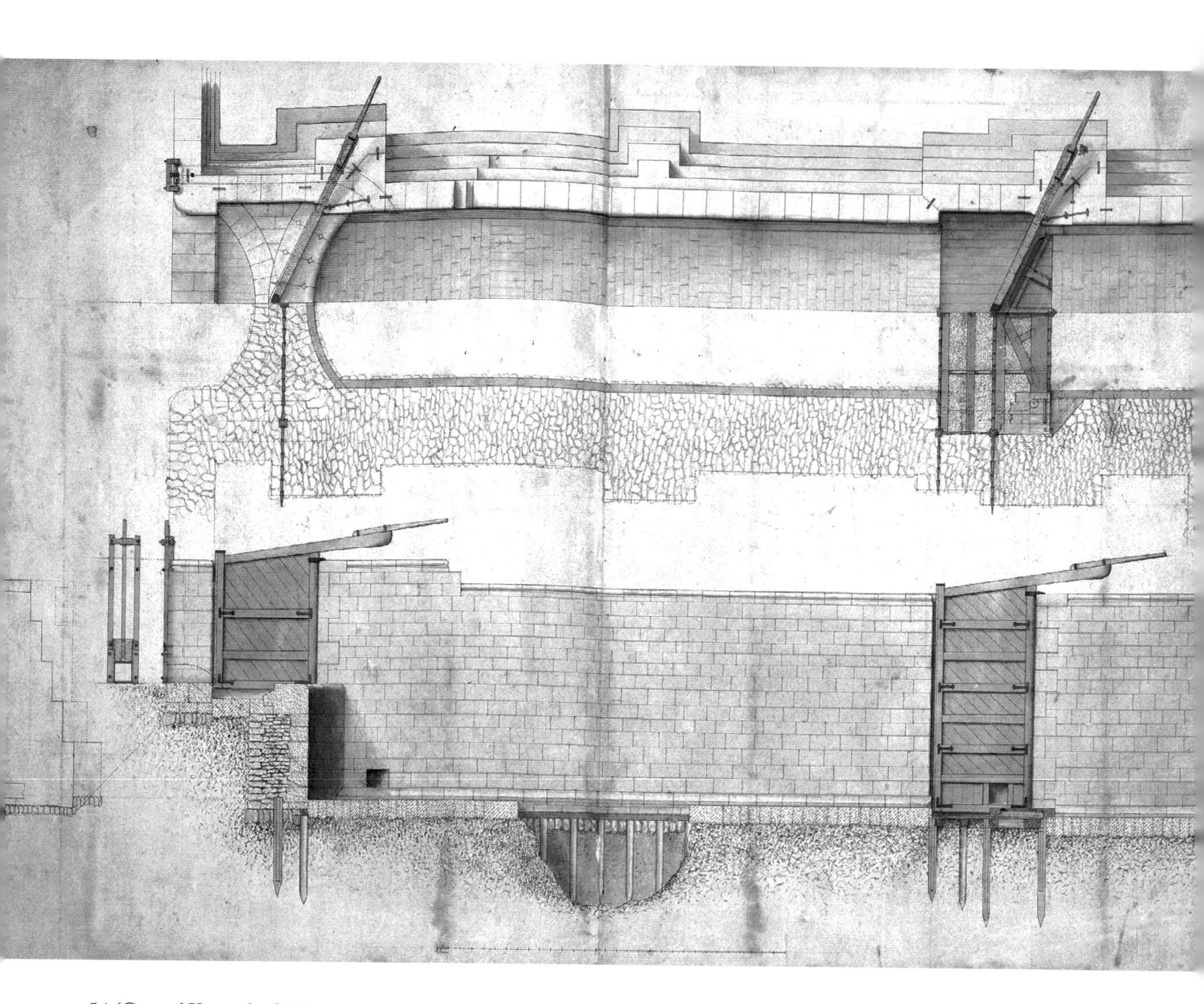

54. 'General View of a Calder Lock', Smeaton's drawing showing the construction of the lock sides, stonework, lock gates, ground and gate paddles, and different types of piling.

1761

Week 1 May 4th - 10th

Wednesday 6th
This Evening arriv'd at Wakefield and took a general View of the Works there.

Thursday 7th, Friday 8th, Saturday 9th
These Days look'd over and examined all the Works which at the several Places appear'd as follows.

At and from the Work Yard the following Machines etc have been produced. Viz.
1st A Floating Stage with Screw Legs, for supporting the Engine, in driving gauge Piles for the Dams; employ'd at present at Lupsit
2nd Three new Piling Engines
3rd Two small Water Screws to go by hand, with Frames etc to D[itt]o.
4th One large D[itt]o and frame, with horse Frame & Stampers, not quite compleated, but going on with all Expedition
5th A Horse Frame for the 2 feet Water Screw ready for Setting up
6th A laving Wheel to go by hand not quite compleated.

Since my last the Platform in the Yard has been remov'd with its Shed to Netherton Quarry; and the Flat or Punt that was sunk near Thorns Lock, has been weighed, and is now in use.

A Quantity of guaged, rabbeted, and groov'd Pile for the sheeting of Lupsit Dam, &Wakefield Lock, Centers for the Road Bridges, Molds for the Stone and other Works, Shades for the Workmen etc etc, has been sent from the Workyard

At Wakefield Lock the Walls are raised above low Watermark, and the Earth made good to the same height, the Engine order'd to cease work after Monday; the Ground whereon this Lock stands is almost the worst possible, and therefore as its lower Gates will shut against 14 feet head of Water an Extraordinary Security will be requir'd in the Floor of the Chamber, which is therefore ordered to be deferr'd till the Walls are up.

The Cut is compleated, all but the Solid left standing against Wakefield Dam the two Road Bridges over the Cut are paved and passable for all Carriages but the Parapets are not compleated, some small Settlements have happened in both these Bridges arising from the necessity of the great Obliquity of the Cut and the Road but which don't appear to have increas'd since the Centers were struck; and are therefore of no great Consequence.

At Thorns Lock and Cut nothing done since my last, nor does there appear any damage by the Flood or otherwise, except the Frost has peel'd off a little of the pointing of the Lock Walls in some places, where the wetness of the preceeding Season had prevented the Mortar from hard'ning

At Lupsit the Bridge over the Cut answerable to the Ford is building, and near 3 feet higher than the intended Dam's Top. The middle row of gage Piles for Lupsit Dam drove, and a temporary Dam inclosing the Ground for the Dams end & Wall, the Conduit of the draw Gates, and 2 Bays of the Dam; the Timber Work for the Dam within this enclosure compleated, the Dam's end and Conduit Walls raised almost to the height of the Dam Top and ready for the Arch; the Conduit Floor compleated; One Man pumps the leakage here.

At Horbury Mill Cut is a terrible Scene of Devastation wholly arising from the extream looseness of the matter where this Cut is dug being almost entirely a Mass of running Sand and Gravel, which the Floods rising here very high have washed down from the Banks into the Cut on one side and upon the Meadow on the other Side, have therefore order'd the foot of the bank next the Cut to be cleared for 4 feet and the matter thrown back; and to commence a fresh Slope so as to be almost double to the former, so that, tho' the Cut will be at least 6 feet wider at top than before yet it will not be wide enough for the passage of two Boats in all places. As these Consequences cou'd not be foreseen when the Cut was set out; nor yet before some progress had been made therein, I apprehend this Method will be the easiest to render this Cut passable and supportable, but if the Committee has any Objection hereto shall do it conformable to their Orders. At the Lock Pit a temporary Dam is made to prevent the Leakage of the Cut from communicating therewith, and an Engine is ready to be fix'd down: the Head Cut is compleated, all but the necessary Solids for keeping out the Water till the Lock is compleated.

At Washinstone the Head Cut is compleated and Lock Pit down to Low Water

At Horbury Bridge the Head Cut is compleated and Lock Pit down to Low Water 3 Gins are driving temporary Piles for shoring the matter while getting down the Wings and Sells, which in those shingly Foundations is exceeding troublesome.

Barnsley Road Quarry worked out & backing for Wakefield Lock and some Aisler for Lupsit Dam, getting at Agbridge Quarry

At Adingforth Quarry a large Quantity of good Aisler, Dam Stones, Wall Stones and Backing has been got and carried to Washingstone and Horbury Mill Locks as well as much remaining in Place. The Dam Stones & Wall Stones got last Year at Hartly bank have been scapel'd up ready for use at Horbury Mill & Washingstone Locks

At Netherton Quarry a large Quantity of excellent Bluestone Blocks for Sells and hollow Posts have been rais'd and work'd nearly sufficient for the use of the present Season besides a large Quantity of excellent Aisler, Damstones, Wall Stones, Backing Quoin Stones Caping etc part of which on account of its excellence has been carried to Wakefield Lock; that is Sell Stones & hollow Posts.

At Millbank Quarry a considerable Quantity of Aisler Damstones Wall Stones and Backing has been raised and work'd and the Quarry appears very promising for these uses, and in Case the long Cut is carried into Execution the Stones may be loaded from this Quarry immediately into Vessels navigating thereupon to serve the Works above

During these Days Messrs Nickals & Smeaton took a fresh level from Wakefield Dam to the Lock and found the Level of Wakefield Dam to be 14 feet above the Lower Gate Sell of the new Lock. Also from Lupsit reach to the tail of Horbury Mill Cut, 3dly from Horbury Mill Dam to the tail of Horbury bridge Cut, and from thence to the figure 3 Cut; from whence it appears the Dam at Lupsit will be 3f 9in. that at Washingstone 3ft 6in & that at Horbury Bridge 6 ft.

1761 Week 2 May 11th - 17th

Wednesday 13th
This Day at Wakefield to look over the Works and in the Afternoon to Halifax

Thursday 14th, Friday 15th
Attending the meeting of Commissioners at Halifax & returning

1761 Week 3 May 18th - 24th

Monday 18th, Tuesday 19th, Wednesday 20th, Thursday 21st, Friday 22nd, Saturday 23rd
This Week attending the Works at Wakefield instead of Mr Nickals then employ'd in making up his Acc[oun]ts during the Course of this Week the hands in the Yard were employed in compleating the Horse Frame for the Engine to be erected at Washingstone Lock making 150 feet of Spouts or Troughs[275] for carry[in]g off the water from the Horbury Mill Lock Pitt; while the Diggers are bottoming the Tail Cut, by Lengths of 50 Yards at a Time, repairing Piling Engines, Pumps, and other necessary works.

At Fall Ing Lock the Forebay & Sell compleated & the whole rais'd 6 feet above Water.

At Lupsit Dam the Hands going on with setting the Apron without the Draw Gate Conduit, setting up the Center and turning the Arch. At the South End an Enclosure being made for digging the foundation for the Dams End Walls, the Diggers were going on therewith, and one of the small Screws placed for draining the Water, but on Wednesday night the whole was interrupted by a fresh coming down, which overtopped the Inclosures and filled the Works; which continued almost the remainder of the Week, during this Interval the workmen employ'd at Lupsit Bridge, and compleating that at Thorns; the Timber work being got on, and the Labourers employ'd in making good the Earth at the Abutments in order to form a Road.

275 Exactly how these troughs were used is not clear, there is a later reference on 8th July to troughs being prepared at Addigforth Quarry, for Washingstone dam.

At Horbury Mill Pasture the Diggers employed upon the Tail Cut as specified in my last but found it necessary to Shift the Banks 5 feet of a Side further back in order to form the necessary Slope, this the Undertaker proposes to do for 7s 8 p[e]r Rood, & 1 shilling p[er] Rood for setting the face with Sods 3 Course high the Length upon measuring turns out upwards of 400 yards. On Saturday the Troughs were fixed for the first 150 Yards. The former part of this week Carpenters employed in fixing the 2 ft Screw and Horse Frame at Horbury Mill Lock Pit which was got to work on Thursday and on Friday the Diggers got in to clear out the Sils and bottom the same, but on Saturday were drove out by the Rain.

At Washingstone 2 Carpenters and Laborers employ'd in driving Piles for circumscribing the matter for the Excavations for the Engine Pit for the Wings of the Stern Sheeting, Stern Bay and forebay, for want of which original Labour those Excavations have from the extream looseness of the matter turn'd out extreamly troublesome. On Saturday the Engine Pit was got down 7 feet below the adjacent Surface of the Calder, the Gravel very open and full of Springs. On friday the Horseframe was bro[ugh]t to place, & the Carpenters begun to fix up the Engine.

At Addingforth and Netherton Quarrys several of the Laborers, taken of to the Piling and other Works and more will be wanted as soon as Horbury Mill and Washingstone Lock Pitts are got down, and ready for Piling, so that till we are fixed with the number of Men that will be required to be drawn from those Quarrys, no bargain can properly be made for getting Stone.

At Netherton the Masons are preparing materials for Horbury Bridge Road Bridge.

Have inspected Doncaster and Barnsley Road Bridges but don't find any great Alteration since I came down, the greatest apparent Defect being in the original setting of one Side, of one of the Arches, and has happen'd as I conjecture by some bearing wedge getting out, or the Settlement of the Ground itself under the Prop under one End of one of the Ribs of the Center when the Arch was nearly at a close which was not perceiv'd by the Workman till after the Center was Struck, this Defect has happen'd in the most unlucky place possible, being in that Side of Barnsley Road Bridge which is seen from Doncaster Road Bridge.

1761 Week 4 May 25th - 31st

Monday 25th, Tuesday 26th, Wednesday 27th
Attending the Calder Works. Hands in the Yard making and grooving Plank and Guage Piling, making Wheel Barrows, sawing Timber for Lupsit Bridge, filing up Piling Engines, and other necessary Works

Wakefield Lock going on Monday and Wednesday as per last, but on Tuesday all Hands off on Acco[un]t of the Rain, as was likewise the Case at all other Places except the Yard, on Wednesday Evening the Walls were between 7 & 8 foot above Water.

On Monday & Wednesday the Hands at work at Lupsit Dam, and compleated the Arch over the Conduit, but on Wednesday the Work being full of Water the Hands chiefly at the Bridge.

At Horbury Mill Pasture the digging going on as before, and on Monday bottoming the Lock Pit; but on Tuesday beat off by the Rain (as at other Places) and on Tuesday night the Lock Pitt was fill'd by the fresh coming down, and rising higher than the temporary Dam, which was not set about to be cleared till Wednesday afternoon, when the Water was sufficiently abated. The Carpenters there making Sheds for the Work Tools, for the Gin Horses & for covering the Engine Wheel.

At Washingstone, Monday and Wednesday get[in]g up the Engine which the Latter Day was almost compleated for work.

At Addingforth &Netherton Quarries but few Hands at work, either Masons or Quarrymen on acc[oun]t of the broken Weather.

On Wednesday Evening Mr Nickals having digested his Accounts ready for Mr Simpson, and Mr Smeaton having rec[eive]d a Letter requesting his Attendance at Durham, to settle a particular point there without which the intended Work cannot proceed, and having left his Directions and Observations with Mr Nickals, set out for Austhorpe this Evening in order to set forward for Durham next Morning proposing to return so as to attend the Calder meeting on the 4th of next Month.

1761 Week 5 June 1st - 7th

Thursday 4th
This Day attended the Meeting of the Comm[ission]ers at Halifax

Friday 5th
This Day overlook'd all the Works which were as follows.

Review'd the Situation of the Long Cut.

At Mill Bank Quarry all the Quarry men discharg'd and gone, & Masons compleating their Bargain of hewing work which wou'd be done this Day.

At Netherton the Quarrymen order'd to proceed no farther with their Bargain than to get what arch and Quoin Stones are necessary for Horbury Bridge Road Bridge; the Masons there to finish their Sells & Hollow posts now in hand, and the Arch and Quoin Stones for the Bridge last mentioned, and then to Stop.

At Addingforth Quarry all the Quarrymen discharg'd and gone, One Mason that had been hurt in the Work compleating his Bargain of Damstones:

At Washingstone Lock, the Engine at Work, and the Diggers getting down the Lock Pitt, being then about 3ft below the Level of the Calder: A Large feeder of Water here the bottom being like the rest, an open Slough; A Bed of running Sand next the Calder at the lock Tail requires to be secur'd with a temporary Fence.

At Horbury Mill Lock 3 Engines employed in driving Piles for the Foundation; the Diggers at Work bottoming the Tail Cut 2ft below the Level of the Calder which being loose & leaky requires to be done by parts as mention'd in a former Report.

At Lupsit Dam the Arch over the Conduit being turned and the North end Wall got out of the reach of the Water the Masons proceeding with setting 2 Bays of the Dam but not quite compleated; at the South End Wall, two Engines proceeding with piling, the Masonry of the Bridge passable for Carriages;

The work at the Head of Thorns Lock having been laid dry by means of one of the small Screws, the Masons were setting the upper Gate Sell and proceeding with the Lock Walls the Labourers ramming & filling behind them.

At Wakefield Lock the Walls rais'd to their intended height wanting the Caping, which with the Setters for the bottom are not yet deliver'd in by the two Contractors, tho' order'd 4 months ago. Some hands at work getting on the Parapets of Doncaster Road Bridge.

In the work Yard framing the timber work of Lupsit bridge, and considerably advanced; the other hands at Work in preparing guaged, grooved, & rabitted Piles

June Saturday 6th
Return'd to Austhorpe, in order to begin upon the Estimate of a Lock & Dam order'd against the next Meeting.

<u>1761 Week 6 June 8th - 14th</u>

Monday 8th, Tuesday 9th, Wednesday 10th
These Days upon the Works at Wakefield etc

Fall Ing Lock standing still as p[er] last for want of Materials from the Contractors... Hands in the Yard preparing Piling for Lupsit Dam Horbury Mill & Washingstone Locks, and finishing the Timber Superstructure for Lupsit Bridge.

At Thorns the Forebay & Sell compleated and the Walls advancing with all diligence being ab[ou]t 8 Feet above Water.

At Lupsitt Dam the Conduit & 2 Bays of Dam setting compleated, at the South End Wall the Foundation Piled and Pitch'd and laying the first Course of Stone, the Carpenters and Pile drivers going on with the Timber Work for 4 Bays so much being inclos'd by the temporary Dam at the South End.

At Horbury Mill Lock the Stern Sheeting compleated the Bearing Piles greatest Part of them drove; Part of the Stern Bay and part of the Forebay Plank Piling dispatch'd; thc Excavation for the Stern Bay inclos'd with temporary Piles to prevent the loose Matter from incessantly falling in; expected to be ready to begin the Pitching this Week; the lowest Stage of the Tail of the Cutt is cleared; but as, till the affair of the long Cutt is determined Mr Platts will have only finishing Work this Season he has thought proper to shorten the Number of his hands, Those which remain are at present employ'd in getting down Washingstone Lock Pitt the Stern Part of which is now bottom'd, and the Piling Work begun therein. This Lock Pitt is extreamly leaky being almost an universal Spring.

At Addingforth the Man mentioned in my last who was hurt at Wakefield Lock, and his Boy preparing by Task Work some Wedge Courses[276] necessary for Washingstone Dam; at Netherton getting and working the Bridge Work by Task.

At Mill Bank all Hands gone, on Tuesday took an exact Level of the Tract of Ground through which the Long Cut will Pass, and found it not materially to differ from my former observations. In doing the Upper part of which from Thornhill Lodge to Dewsbury low Ford Mr Wilcock[277] was present. The Turnpike People are again at Work getting Gravel at Lupsit Ford, but so far below the Dam as not to affect it.

<u>**1761 Week 7 June 15th - 21st**</u>

Monday 15th, Tuesday 16th

Employ'd in compleating the Estimates for a Lock and Dam upon the Calder, and making a Scetch of the Grounds thro' which the propos'd Long Cutt ought to pass in order to be laid before the Commissioners at their Annual Meeting on Thursday next.

Wednesday 17th

This Day went over the Works and from thence to Halifax.

the Walls of Thorns Lock being got up to the Caping, the Hands employ'd in getting out the Water for Setting the Bottom.

---

276 Andy Beecroft suggests that 'wedge courses' would be wedge shaped masonry necessary to provide the curvature of the dam. This can be seen on the drawing for Lupset Dam, (RS JS/6/46) which shows specific courses cut to a wedge shape as indicated by the hatched lines.

277 William Wilcock was asked to survey and value the lands at Wakefield, Lupset and Horbury, and negotiate with the landowners to buy the land. (WYJS/CA MIC2/4 Letter to Mr Wilcock January 1760)

Thursday 18th
This Day at Halifax attending the Meeting of the Commissioners.

Friday 19th, Saturday 20th
These Days upon the Works the State of which are as follows.

In the Work Yard the hands employ'd upon the Piling and on the Lock Gates for Thorns Lock.

Fall Ing Lock standing still for want of caping & setting as p[er] last,

Thorn's Lock when the Water was got out a large Quantity of Mud and some Gravel that had not been clear'd out the last year; this being clear'd out the Masons at Work in setting the Stern Apron; the Eddystone composition of Watchet Lime and Pozolana with which the Sell Pieces and Hollow Post Stones were united which have been under Water all Winter was prov'd and found to be fully equal in Hardness to that of the Stone.

At Lupsit Dam the Staple Posts and head Pieces for the Gates were got in and also a Set of Chain Bars[278] for uniting the Dam's End and Conduitt Walls together, the Conduitt being now ready for turning the Current of the Calder thro' it. The South End Wall got up above the height of the Dam the Apron pitched; The Timber Work for 3 Bays (in my last Journal by mistake call'd 4) compleated and the setting of these 3 Bays going on, the Carpenters at Work in driving the Heart Plank Piling for another inclosure;

at Horb[ur]y Mill Pasture the Diggers bottoming the Tail Cut; The small Screw being now work'd by a small horse & same is enabled to take a quarter Length at once, at the Lock all the Pile Work drove except a Breast Work for the Stern Recess as an additional Security in those deep Locks in open gravel and quick Springs; the rough Pitching and the String Pieces on the bearing Pile heads under the Lock Walls completed, the Excavation for the Stern Bay also done. The Horse Stampers also for the Lock ready for Use.

At Washingstone all the bearing Piles drove and the Masons rough Pitching under the Walls, the Stern Sheeting and that under the Forebay compleated, that for the Stern Bay going on; the Leakage of this Lock is nearly a Match of the Great Screw work'd by 2 Horses. All the Hands were let know that it was the Pleasure of the board that no Working wou'd be continued of more than six Days a Week; except in Cases of Necessity, and therefore all such as were not at Work where pumping or drainage of Water was going on were to work 12 hours a Day.[279]

---

278 Andy Beecroft suggests that chain bars would be some form of reinforcement. Perhaps a heavy chain was laid in a masonry course and bars fixed vertically at intervals through the links to tie the structures together.

279 meeting of the Commissioners 18th June 1761, (WYJS/CA MIC2/2)

1761 Week 8 June 22nd - 28th

Wednesday 24th, Thursday 25th, Friday 26th
These Days upon the Works.

At Fall Ing Lock some of the Capings d[elivere]d in but Widow Hartley not being able to deliver any more at the Price contracted for, that Work still stationary for want thereof.

In the Work Yard preparing Pile Work for Horb[ur]y Mill & Washingstone Locks and also for Lupsit and Washingstone Dams and at the Gates for Thorns Lock,

At Thorn's Lock setting the bottom of the Chamber putting on the Capings and regulating the Faces of the Sells and hollow Posts.

At Lupsit Dam the 3 Bays at the South End being got out of the way of the Water the Masons compleating the same & having formed a new Inclosure for 4 Bays more the Pile work therein going on.

At Horbury Mill the Masons rough pitching the Foundation of the Forebay & Side Walls and the Excavation making for the Stern Bay which being enclos'd with a temporary set of rabbitted Piles the leakage therein is great reduced.

On Wednesday the first Guage Pile of Washingstone Dam was drove and the Workmen proceeding therewith.

At Addingforth 4 Men preparing Wedge Courses for Washingstone Dam.

At Netherton the Arch and Quoin Stones for Horbury Road Bridge being most of'em prepared the Masons call'd off to other Works, no Hands at Work there.

1761 Week 9 June 29th - July 5th

Wen'sday 1st, Thursday 2nd, Friday 3rd
These Days upon the Works.

At Wakefield Lock 2 or 3 Hands preparing the Capings sent in last Week by Widow Hartley.

In the Yard all Hands at work upon the Piling the Lock Gates being suspended on Acc[oun]t of the quantity of Piles wanted at Lupsit & Washingstone Dams.

At Thorn's Lock the Masonry compleated and the Labourers at Work in clearing away the temporary Dam and Solid left at the Lock Tail.

At Lupsit Dam the End Walls got up to the square ready for the caping the 3 South Bays compleated, the Masons and Carpenters at Work upon the 4 Inclosed Bays.

At Horbury Mill Pasture Mr Platts Men bottoming the Tail Cut; and the Masons at work in the Lock Pit Laying down the Stern Bay and Sell, the first stone of which was laid down on Wednesday, and the last on Friday and on Saturday is expected to be ready to commence the measur'd Work.[280]

At Washingstone Lock the Masons and Carpenters rough pitching and compleating the Foundations for the Stern Bay.

At Washingstone Dam all the Heart Guage Piles driven and 6 Bays of Plank Piling therein; the temporary Dams made for inclosing the South End Wall, Conduitt & 6 Bays of the Dam and the Bearing Piles driven for the Front of the South End Wall.

Adingforth Quarry some Masons at Work preparing necessary Materials for Washingstone Dam and Lock.

At Netherton & Mill Bank nobody at Work

1761 Week 10 July 6th - 12th
Wednesday 8th, Thursday 9th, Friday 10th

These days at Wakefield, Halifax and on the Works

At Fall Ing Lock all the Capings faced that were bro[ugh]t in no Hands at Work there.

In the Yard the Hands at the Piling and Thorn's Lock Gates The Rain on Monday having produced a Fresh on Tuesday Morning broke over the Dam into the Tail of Thorns Lock and put off the Hands employ'd there.

At Lupsit Dam the Works interrupted by the Fresh and some of the Corn Mold Earth wash'd out of the temporary Dam; but those repaired and 5 Bays on the South End compleated and the Carpenters Work for 2 more ready for setting; the Carpenters and Pile drivers preparing for making a Shut of the last 4 Bays.

At Horbury Mill Pasture the Diggers in the Tail Cut also interrupted by the Fresh but again at work as before. In the Lock Pitt no Interruption, The Masons on Friday had compleated the Aisler of the Chamfer'd Course and most of the backing up to the square.

---

280 Andy Beecroft suggests that it may well have been agreed that work up to a certain level was done on a daywork basis due to the awkward nature of working at lower level. Above this more routine construction like sill walling etc would be done on measure to ensure productivity. As measured work would be the norm mention of it may be infrequent.

At Washingstone Lock the Masons pitching the Forebay the Stern Bay being compleated for the Mortar Courses. And this Week will be ready for the Masonry.

At Washingstone Dam some hindrance by the Fresh but no damage, of the heart Plank Piling 8 Bays compleated and 4 d[itt]o of the Skirt, the Inclosure for the South End Wall & Conduit turn'd and so leaky as to employ 2 Pumps, 2 Men at each, but one of the hand Screws being bro[ugh]t up from Thorns Locks Tail. 2 Men kept out the Water with the greatest ease.

At Adingforth Quarry working Troughs & Wedge Courses for Washingstone Dam.

At Netherton 2 or 3 Men at work making Sell Pieces and Staple Posts for the Conduit Cloughs of Washingstone Dam.

1761 Week 11 July 13th - 19th

Wednesday 15th, Thursday 16th, Friday 17th
These Days upon the Works & resurveying the Ground for the Cut of Dewsbury Mill Mirfield Low Mill and Ledger Mill and also the Grounds for a long Cut between Ledger Mill & the Boat house mention'd while the Act was obtaining in order to render the Navigation independent of Mirfield Low Mill, being in length about 7 Furlongs.

At Wakefield Lock no Hands – some Hands employ'd in putting on the Parapet of Barnsley Road Bridge. In the Yard chiefly at the Piling and some Hands at Thorn's Lock Gates; At Thorn's Lock no Hands. Mr Platts Men at work upon the Solid left between the Rocky & the clear Part of Lupsit Cut.

At Lupsit Dam all the Masonry compleated except the last 4 Bays and those shut up with the Plank Piling both in the heart & Skirt, and the temporary Dams compleated, so that now the whole River goes through the Conduit Clough which wou'd run off 3 times the Quantity with 2 Feet of [space in text]

The Diggers Work at the Tail Cut of Horbury Mill Lock without the temporary Dam compleated and the Engine removed. At Horbury Mill Lock one side of the Chamber above Water and the other 2 Courses above the chamfer'd Course, but the Forebay & Head Walls want raising from the setting.

At Washingstone Lock the Stern Bay and Sell nearly compleated the Forebay begun, the chamber & stern Walls one course above the Chamfer but the Head Walls rais'd from the Pitching.

At Washingstone Dam the South End Wall 2 Course high and also the Conduitt Wall the Sell laid and setting the Staple Posts in the Walls Ten Bays of heart Piling and 7 of Skirt Piling drove with all the Guage Piles for both the Inclosures takes in 7 Bays of the South End of the Dam.

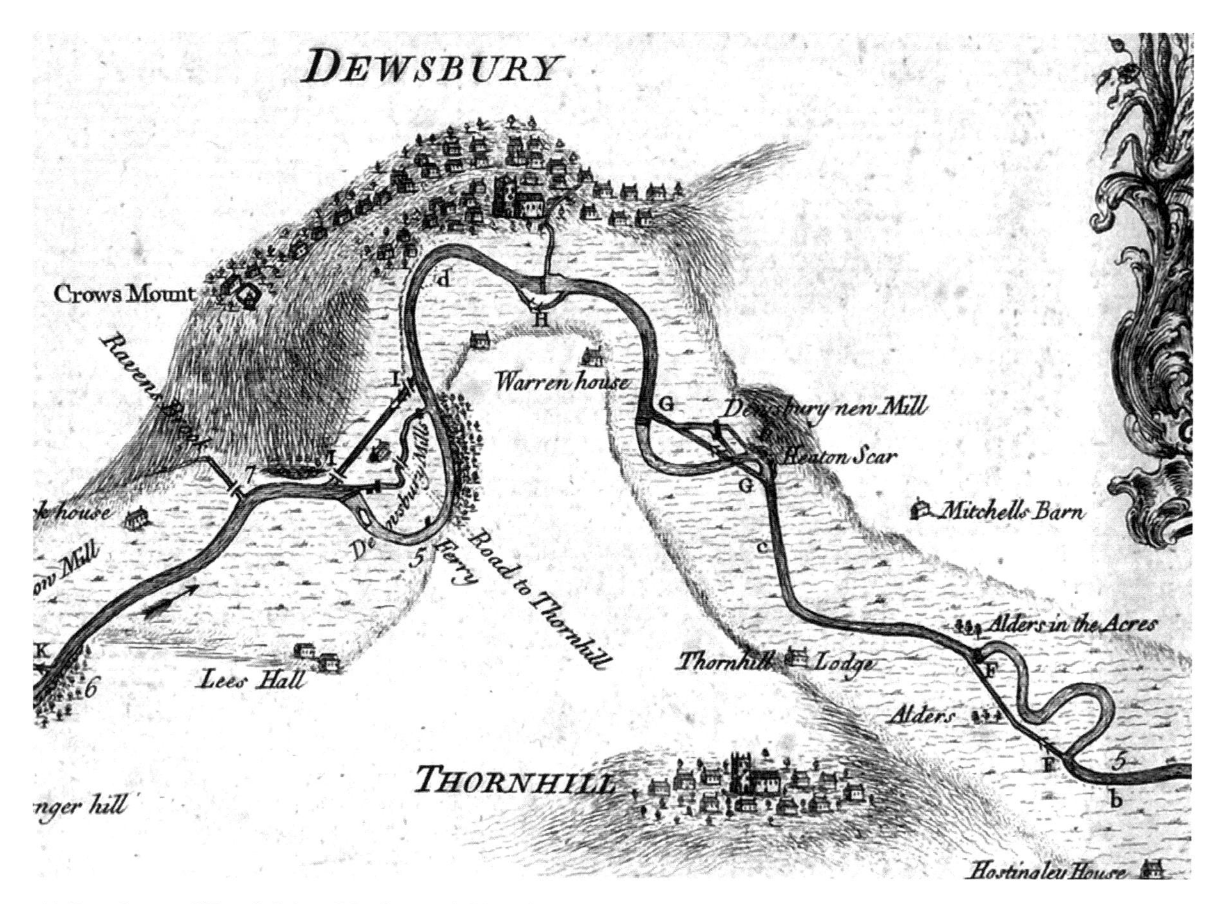

55. Dewsbury and Thornhill from John Smeaton's Plan of 1757/58

At Adingforth some Hands at work making Troughs for the Lock and the necessary Pieces for the Dam.

At Horbury Bridge Lock Pitt setting up the little Engine bro[ugh]t from Horbury Brid Mill Cutt Tail and got it to work ready for the Diggers.

At Netherton no hands.

<u>1761 Week 12 July 20th - 26th</u>

Wednesday 22nd, Thursday 23rd, Friday 24th
These Days upon the Works and staking out the Figure of 3 Lock and long Cut from thence to Dewsbury which is now ready for the inspection of whoever may be appointed to fix the Value thereof

At the work Yard at the Plank Piling and Lock Gates, the Diggers at work on the solid in Thorns Cutt.

At Lupsit Dam one of the 4 Bays contain'd in the last Inclosure compleated and the Skirt Stones and upper Springers[281] of the 3 remaining set with some Courses of Setters so as to render the whole above Water.

At Horbury Mill Lock the Chamber and Stern Walls 6 Courses above the Chamfer and the Forebay and head Walls raising to the the same height.

At Washingstone Lock the Walls 4 Course above the Chamfer and setting the Upper Gate Sell; at the Dam the South End and Conduit Walls 4 Course above the Floor and setting the Conduit Floor; the Heart Piling shut in quite across the River, and the Skirt Piling wanting only one Bay, the Carpenters fixing the String Pieces Tye Beams and Heart sheeting of the 7 Bays within the Inclosure, the Ground of all the Bays without the Inclosure secure from Blowing by a Quantity of Rubble Stones filled between.

At Horb[ur]y Bridge Lock the Diggers at work bottoming the same, this Lock Pit is likely to turn out the dryest we have yet had, but the Ground still the same viz. an open Gravel.

At the Quarries no Hands at work except those concerned in leading Stone.

Considering the Openess of the Ground of the Calder Valley wherever we have penetrated into the same have contriv'd a false Apron to be applyed to the Stern Gate Sell of the deep Locks, and have directed Mr Nickalls to proceed with the same.

<u>**1761 Week 13 July 27th - August 2nd**</u>

Wednesday 29th, Thursday 30th, Friday 31st, Saturday August 1st
These Days upon the Works and resurveying the Calder from Batty's Mill to Brooksmouth.

The Hands in the Yard employ'd upon the Piling for Horbury Bridge Lock and upon the Lock Gates.

At Lupsit Dam the Work being clos'd on Tuesday The Labourers employed on Wednesday in drawing the Piles and removing the remainder of the temporary Dams, Wednesday afternoon being very rainy everything cleared for a fresh, as well here as at the other works: on Thursday Morning a fresh came down and besides what was discharged by the Cloughs run over the Dam 20 Inches thick upon the Crown, on Friday Morning all the River again taken by the cloughs and

281 springer - the point or place where the curve of an arch or vault begins.

9 Inches within the Dam cou'd perceive no Damage, ordered the Cloughs to be put in in order to regulate the Rubble Stones that had been thrown in at the Foot of the Conduit Apron.

The Masons employed in making Wharf Walls of dry Rubble for supporting the Banks going in and out of the two Bridges of Thorns & Lupsit, the Diggers at work on the Solid at the end of the Rock in Thorns Cutt;

At Horb[ur]y Mill Lock the Forebay compleated and the Sell set, the Walls of the Chamber at high water mark; the Fresh filled the Lock Pitt but no damage or hindrance to the work;

At Washingstone Lock the forebay likewise compleated & Sell set, the walls of the Chamber being near upon high water Mark, the Lock Pitt fill'd by the Fresh but no damage or hindrance to the work. At the Dam on Wednesday the Workmen were putting on the Lidds of the Conduit and setting the Skirt Stones and upper Springers, but the fresh put out all the workmen here; Yet without material Damage;

At Horb[ur]y Bridge Lock the fresh put out all the Diggers, being within one Day of compleating their work for the Piling, but on Friday the Engine again got to work to get the water out.

Mr Platts wanting work for his Men set them upon the figure 3 Lock Pitt which was begun on Thursday 30th

<u>**1761 Week 14 August 3rd - 9th**</u>

Wednesday 5th, Thursday 6th, Friday 7th, Saturday 8th
In this Week has happened one of the most mortifying Accidents, that I hope ever will happen to these works,[282] of which I gave the Committee an acc[oun]t in a Letter from Wakefield last Friday and which upon further Inspection does not appear to be worse than then represented, the Water being more down on Saturday (tho' still running over the Heart Planking which terminates 10 Inches below the Crown) The Aprons and Skirt was particularly examined and found to be entire, have concerted Measures for repairing and securing the present; and also for avoiding the like Misfortune in the others.

Hands in the Yard employ'd upon the Gates for Thorns Lock being all fram'd and planked and upon the Piling for Horb[ur]y Bridge Lock Pitt the Minion being all ground for this Year's Service the Engine removed from Wakefield Lock & set up at Horb[ur]y Bridge, the Tail Cutt at Wakefield Lock was set out for the Diggers.
The Diggers in Thorn's Cut as per last,

---

282 About 20 yards of the crown of Lupset Dam were lost due to some of the stones on the crown coming loose. (Ref: WYJS/CA MIC2/4 letter to the Commissioners 7th August 1761)

Horb[ur]y Mill Lock got up to the Caping and began to set the fore Apron and back up the Walls with Earth & Gravel.

Washingstone Lock up to the Caping and begun to set the Chamber and Stern Apron.

Washingstone Dam nothing done since last Week the Earth of the temporary dams being no sooner repaired but were attack'd by a fresh flood. On Thursday no Damage here of consequence. On Saturday began to repair the Earth was[he]d out of the temporary Dams.

At Horbury Bridge the Diggers had not been able to compleat the Lock Pitt before the fresh Flood came on, The Drainage there being considerably increase'd by the Rains and floods this Interval was made use of in getting up the great Engine from Wakefield Lock and was got to work on Saturday; the Piling will go on with all possible Expedition as soon as the Digging is done, but I apprehend cannot be completed in a fortnight as none of the Former have been done in that time tho' free from Interruption.

At the Figure 3 the Diggers at work upon the Lock Pitt, have order'd the Banks to be well Sodded next the River to the height of the highest Floods.

<u>1761 Week 15 August 10th - 16th</u>

Monday 10th, Tuesday 11th, Wednesday 12th, Thursday 13th, Friday 14th, Saturday 15th
Attending the Works;

On Monday the Water at Lupsit Dam being rained from the Breaches a Pair of Carpenters were set on to fix in a Crown Timber, in order that the Crown Stones may be fixed down thereby, and 4 Labourers employ'd in recovering the Stones from the Dams Skirt. On Tuesday D[itt]o. On Wednesday an Additional Pair of Carpenters were set on at the other end and this Day all the Stones were recover'd. On Thursday the Carpenters having compleated some of the Bays the Masons begun to set. On Friday hard Rain in the Forenoon beat off all the workmen and in the afternoon came down a fresh which rose about ? yard above the Crown[in]gs the Dam which in the Evening was somewhat abated cou'd not observe that anything was displaced since the Workmen left it except that some loose Stones which from the fineness of the Preceding Day, the Workmen had laid upon the Skirt to be ready were again swept off from thence. In the night more rain so that on Saturday Morning the Water was again mended & at Noon rather abated when everything appeared in Place & likely to continue.

The State of the other Works as follows.

In the Yard preparing Piling for Horbury Bridge Lock and Dam, and at the Gates for Wakefield Lock also making Patterns for the Gate Collars and for the Rack and Nuts for drawing the Cloughs.

Diggers at the Tail Cutt of Wakef[iel]d Lock, at the Solid in Thorns Cut and getting out some Clay that had been left with a view to make brick in case they had been wanted for backing etc. On Wednesday Morning Mr Thompson the Surveyor of the Road leading from Wakefield to Dewsbury had set on a number of Carts to lead away the Gravel that had been thrown up by the Dam during the late Flood, and which by Penning the Water 6 or 8 Inches upon the Apron thereof I look'd upon as an additional Security. I therefore desired and advis'd him to desist, he shew'd me his Act of Parliament[283] which allowed him to take Gravel within 15 Y[ar]ds of any Dam, but I told a later Act had said that if any Person did any Act to the Prejudice of the Calder Navigation it was made Felony and therefore advis'd him to take care, he answered that he shou'd acquaint some of the Gent[leme]n and mention'd Messrs Milnes,[284] he continued at Work that Day but have not seen him since.

At Horbury Mill Lock the Fore Apron Chamber, Stern Recess and part of the Stern Apron set but not compleated on Acc[oun]t of the Fresh on Friday, the Head flue Walls[285] compleated and the Stern D[itt]o adoing.

At Washingstone Lock the Bottom compleated the Stern Flue Walls in hand.

At the Dam the Carpenters employ'd in removing some part of the upper half Tree in order to enlarge the Base of the Dam and thereby diminish the Obliquity. The Masons upon the South end Wall and Conduit but the whole interrupted by the rain and Fresh on Friday.

At Horbury Bridge Lock 2 Engines at Work upon the Piling and 1 D[itt]o at the Dam, all the Heart Guage Piles drove and the Planks Piling begun but the Work here all interrupted by the Fresh the same as the others. At the Figure 3 the Lock Pitt going on.

<u>**1761 Week 16 August 17th – 23rd**</u>

Monday Augt 17th to Saturday Augt 22nd
These Days attending the Works in the Yard at the Piling for Horbury Bridge Lock & Dam and all Fall Ing Lock Gates.

At Fall Ing the Diggers getting out the Tail Cut

At Lupsit Dam on Monday the Water not being waned from the Breach; nothing cou'd be done till Tuesday Morning when the Hands again begun, not a Stone being moved out of Place by the fresh but the Carpenters not having compleated their Work only one Pair of Setters cou'd

283 Act of Parliament 14 George II, c. 19, 1740 *An Act for repairing the roads from Red-house near Doncaster... to Halifax*
284 Several members of the Milnes family were active as Trustees of the Turnpike being constructed from Wakefield to Halifax.
285 The term 'flue walls' has not been used previously and may refer to the curved walls set on the apron at the entry and exit to the lock.

be employed; Wednesday 2 Pair of Setters employ'd and on Thursday 3 Pair of D[itt]o and having got the Courses in begun to set the Crown, Friday. 2 Pair of Setters upon the Crown. Saturday D[itt]o this ev'ning the Dam's Crown being within one Days Work of being clos'd in to the former Work proposed to the hands to do it on Sunday but they not being all agreeable endeavour'd to cover and secure the vacant Places in the best manner I cou'd.

At Horb[ur]y Mill Pasture Lock all compleated home to the Caping, except some of the Earth wanting behind the Walls. The Engine knocked off.

At Washingstone Lock all compleated home to the Caping except the Head of the Walls which cou'd not be done till the Engine was removed, at the Dam 4 or 5 Bays of setting compleated and going on with all expedition.

At Horbury Bridge Lock, Piling the foundation, which was interrupted by the late Fresh; at the Bridge the Foundation Piling and the Masonry advancing; the Piling work of the Dam there going on with all expedition

The Figure of 3 Lock Pitt going on as p[er] last.

<u>1761 Week 17 August 24th - 30th</u>

From Monday 24th to Saturday 29th
These Days attending the Works:

In the Yard at the Piling for Horb[ur]y Bridge Lock & Dam, and at Fall Ing Lock Gates the Diggers at work upon the Tail Cut and Masons upon the Caping.

At Lupsit Dam on Monday 2 Pair of Setters upon the Crown, and nearly compleated the Original Breach, but in order that the whole might be fixed upon the same principle, this Day took up the Crown of the Standing Part on the South End & the Carpenters begun to put in the Crown Timbers found all the Crown so taken up quite firm & solid. Tuesday the Masons proceeding with the Crown as before and this Day the Standing Part of the North End was taken up and found quite solid like the former, Wednesday D[itt]o and the Carpenters compleated the Crown Timbers Thursday D[itt]o Friday D[itt]o and clos'd in the whole. Saturday the Masons compleating the flue Walls of the Dam and Bridges, and letting in Cramps[286] to the Caping of the Conduit Wall etc. the Labourers employ'd this Day as at spare times before in backing the Dam with Rubble and lining the Skirt with the same.

At Horbury Mill Pasture Lock Labourers backing the Walls with Earth & levelling the Area.

---

286 cramp - a metal bar with bent ends for holding masonry together.

At Washingstone Lock compleating the head flue Walls, and getting on the Caping The Diggers at Work upon the Tail Cut, at the Dam the Conduit & South End Wall compleated & 6 Bays of Setting and proceeding with the rest.

At Horbury Bridge the Arch expected to be clos'd in on Saturday, the Piling work of the Lock and Dam considerably advanced and the Masons begun to pitch the foundations.

During the Course of this Week the Turnpike People employed in carrying Gravel away from Lupsit Ford. Also had a Meeting with the Horbury People concerning the Bridge of communication in the Mill Pasture shewed them the Place proposed by me viz. over the Tail of the Lock, this they objected to because it wou'd cause them to go 10 Yards about and therefore desir'd it to be over the Tail Cut which wou'd make an addition of Expence to the Navigation of at least 100 Pounds at last they agreed to have the Bridge over the Locks Tail as propos'd by me provided they had a communicating Passage by the side of the Cut of 4 Yards Wide and the Bridge over of the Lock of 3 Y[ar]ds wide to the former I agreed but told them I cou'd not make the Bridge wider than those of Thorns and Lupsit without an order from the Committee on account of the precedent those being no more than 10 feet and very sufficient for Cattle & Carriages

1761 Week 18 August 31st - September 6th

Monday 31st, Tuesday Sep 1st
Employed upon the General Estimates for compleating the Navigation on the Present Plan

Wen'sday 2nd, Thursday 3rd, Friday 4th
At Halifax attending the Meetings. Attended Messrs Thompson and Taylor in viewing the Ground of the long Cut[287] who desired some Days consideration in order to communicate their Sentiments to the Committee in Writing

Saturday 5th
This Day upon the Works being hard rain last night this Morning a Flood came down and put a Stop to the Works at Horbury Bridge Dam & Lock. The Masonry of the Dam end Wall & conduit begun, the Arch of the Bridge compleated and the Wing Walls in great forwardness; the Piling and Pitching of the Lock all done except for the lower Gate Sell which if not interrupted by the Flood was expected to have been completed this Day.

The Dam at Washingstone advancing with the North End but this Day the Water over the same, the Diggers employ'd at Wakefield getting out the Solid at the Lock Head the Masons at Work setting the Bottom but interrupted by the Flood; The Hands in the yard at the Piling

287 letter to John Smeaton from John Taylor and Thomas Thompson who were valuing the land at Mill Bank, part of Sir George Savile's estate, for the Commissioners. (Ref: WYJS/CA MIC2/4, 24th August 1761)

for Horbury Bridge Dam and at Wakefield Lock Gates; this Day at 3 oClock the Flood was at the highest when the Water was 3f 3i upon the Crown of Lupsit Dam that is within 3 Inches of the same height as when the Crown was taken off and observ'd that when the Water was at the highest it went over the Dam much more quietly than when 1f 9i upon the Crown as it was at 11 oC[loc]k in the forenoon at 6 oC[loc]k ev[e]ning the Water was abated ab[ou]t a ? Yard, everything then appeared right.

1761 Week 19 September 7th - 13th

Monday 7th
This Day at Wakefield and inspected the Dam at Lupsit; the Water being much abated but still run over notwithstanding the Cloughs were drawn; observ'd everything quite entire.

Wednesday 9th, Thursday 10th, Friday 11th
These Days upon the Works.

At Fall Ing Lock the Hands employ'd in setting the Stern Apron and putting on a Fore Apron to the Stern Bay by way of Additional Security, on Account of the extream Springyness of the Ground whereon this Lock stands; letting in the Boxes for the Gudgeons of the Gates, Straightening the hollow Posts and Sells, letting in the Dowels, clearing the Ground. Piling the foundation and going on with the Stern flue Walls;

In the Yard at the Piling for Horbury Bridge Dam and at Wakefield Lock Gates, at Washingstone the late Flood having taken away the Ground where the North End Wall of the Dam was intended to be plac'd and destroyed the temporary Dams, as soon as the Water was wained off the Hands there employ'd in securing the Ground and constructing a new temporary Dam of Rubble Stone. This accident may be accounted an advantage as it gives Room to make the Dam so much the longer; no material Damage happened to the Part that was completed.

At Horbury Bridge Lock the Masons laying down the Stern Sell, the Sides and Wings being all Pitch'd ready for the Walls; the Bridge backing up with Earth and the Road raising; the Conduit Walls of the Dam rais'd 2 ft high and the Setting of the Floor and Apron thereof completed; Several Bays of the Dam ready for Setting.

Mr Platts at work upon the Figure 3 Lock

1761 Week 20 September 14th - 20th

Tuesday 15th, as Saturday 19th
These Days attending the Works.

The frequent Showers this Week which bro[ugh]t down Spattles[288] of Water every Day greatly obstructed the Works notwithstanding at Wakefield Lock this most difficult and troublesome Bottom was compleated and the Boxes for the Gates feet leaded in; as also the Collars for the Top of the turning Posts and everything prepared for changing the Stern Gates which were bro[ugh]t down to place on Saturday, but the fresh which came down on Saturday Night tho' it did not break into the Lock Pitt, yet by flowing higher than the Engine Spouts oblig'd it to stand and in 12 Hours the Lock & Tail Cut was above 3 Feet deep by Water from the Springs therein; the tail Flue Walls were also forwarded as much as [sentence not completed]

In the Yard at the Piling for Horbury Bridge Dam compleating the Stern Gates for Wakefield Lock and at work upon the Head Gates and also begun and going on with Horbury Mill and Washingtone Lock Gates.

At Washingstone Dam the continued Fresh being more than cou'd be discharg'd by the Cloughs constantly ran over the temporary Dam so that nothing cou'd be done towards compleating the Piling of the foundation, the Water setting through the Cloughs having Pooled a hole near the flue Wall built of dry Rubble, and part of it trusted without Piling at the foot begun to give way where unpiled and was secured by throwing in fresh rubble which shews how necessary it is not to omit anything tending to security upon the River Calder tho' in things of the least consequence.

At Horbury Bridge the Lock Chamber going on with the 2d Course above the Chamfer Pitching & beginning the work for the Forebay and head. The Bridge as p[e]r last. the Conduit of the Dam with its Floor and Arches compleated, the Timber work was interrupted by the frequent Spattles filling and taking down the temporary Dams several times. This Week it is nearly in the same state as p[e]r last, viz the Piling of several Bays compleated, but wants part of the tye Beams which by mistake was not mention'd last Week as being necessary for the Setting. The Diggers at the long Cut and set out the Ground in the low Part thereof in order to its being proceeded upon before winter.

<u>**1761 Week 21 September 21st - 27th**</u>

Tues 22nd, Wed 23rd, Thursday 24th, Saturday 26th
These Days attending the Works which were greatly interrupted by the Spattles coming down the Calder, particularly on Monday and Wednesday Nights, which fill'd all the works.

This Week the Stern Gates of Wakefield Lock were hung so as to turn in place and the Water being got out again on Saturday the Carpenters employ'd in fitting them together. The Diggers clearing the Tail Cutt and the Mason at work upon the flue Walls which in this Lock require to be carried out from the Locks Tail to the Calder, on account of the looseness of the Matter and great depth of the Cutt; The rack & Nuts for drawing the Head Cloughs and the Cloughs themselves compleated.

---

288 spattle - possibly linked to spatter or spittle, i.e small amounts.

In the Yard the Carpenters at work upon the Piling for the Dams, making the Head Gates for Wakefield Lock, compleated the Framing thereof, and took the same in Pieces in order to be carried down to the Lock, other Hands at work upon the Tail Bridge for D[itt]o at the Lock Gates for Horbury Mill & Washingstone.

At Washingstone little or nothing done at the Dam this Week on account of the continued Spattles.

At Horbury Bridge the Lock advancing with all the expedition the Floods will give leave, the Walls of the Tail and Chamber above the Calder's level. At the Bridge the Road raising with Earth and Gravel in order to form the necessary Slope in which a particular Case has occurred which I shall particularly mention in my Letter accompanying this Journal[289]. At the Dam the Masons raising the north end wall to its proper height, and settling the Skirt Stones, but considering the present uncertainty of the Weather and Advance of the Season it was the unanimous opinion of Messrs Smeaton Nickalls and Wilson, not to proceed with the Body of the Dam this winter; but to secure the Piling with Rubble Stone in the best Manner Possible at least to suspend the same till the Weather shall appear favourable.

The Cut above Figure 3 Lock being shallow advances apace.

1761 Week 22 September 28th - October 4th

Wednesday 30th, Thursday 1st, Friday 2nd
Attending the Works.

On Wednesday Ev'ning the Bottom of Wakefield Lock being compleated together with the Flue Walls above the Calder's Level and Stern Gates hung on Thursday Mourning the Labourers begun to take down the temporary Dam; and this Day the same was compleated and the water let into the Lock and the same Ev'ning the Head Gate was got into place ready for hanging, but on Friday Morning on tiyeing the same with the Collar, the Iron proving unsound immediately gave way which stopp'd further proceeding therewith till it shou'd be repaired; This Collar was made of hammer'd Iron with a view to prevent such accidents with the Cast Iron Ones.

In the Yard the Hands employ'd upon the tail Bridge for Wakefield Lock; at Horbury Mill and Washingstone Lock Gates.

At Thorns some Hands employ'd in repairing the Tail Dam raising a Breast Work at the Head of the Lock to prevent the Water overflowing in high floods, letting in Cramps and scapelling stone for the Stern flue Walls.

289 The letter referred to is not included in the collection of letters (WYJS/Calderdale MIC2/4), nor is it mentioned in the minutes of the Commissioners' meetings. (WYJS/Calderdale MIC2/2)

At Washingstone the Water having abated so as to be taken by the Cloughs, the Workmen proceeding with the Piling for the foundation of the Part uncompleated.

Horbury Bridge Lock above Water and getting on the Foresell. The Bridge and Dam as p[er] last, The Diggers at work in the low Part of the long Cut, and ramming the Bank in order to pen the Water 2 Feet above the Grass.

1761 Week 23 October 5th - 11th

Tuesday 6th, Wednesday 7th
These Days attending the Works.

On Tuesday the Water was turned into the Head Cutt at Wakefield and the Lock Gates try'd with the Water upon them, the Head Gate was found to work sufficiently easy and likely in all Respects to answer; and in like manner all the rest of the Machinery; but finding it necessary to lay down an underground Trunk for drainage of the Adjacent Lands as projected last year, the Water was turned off again on Wednesday Morning and this opportunity taken of remedying some small Defects in the Joints between Wood and Stone of the Head Gate, this Day putting on the Tail Bridge over the Lock for the Footway and ha[u]ling Horses.

Workmen in the Yard at the Horbury Mill and Washingstone Lock Gates together with some Plank Piling for the Dams.

At Thorns the Tail Dam compleated the Engine bro[ugh]t from Wakefield hither; the Water out and Diggers at work in clearing the Tail Cut. Masons scapelling Stone for the flue Walls and letting in the Gate Collars.

The Dam at Washingstone again interrupted by the Fresh but the Piles being all driven for the Body of the Dam the Workmen throwing in Rubble for securing thereof; the temporary Dam of Rubble which shuts up the 5 Bays uncompleated, remains unburst and Pens the Water within 9 Inches of its proper height so that the Reach above is rendered Navigable thereby;

At Horbury Bridge Part of the Lock up to the Caping, the Bridge & Dam as p[er] last, the long Cut advancing thro[ugh] the low Ground.

The Latter Part of the Week attending the Quarter Sessions at Leeds but the Bridge Surveyors not there nor anything said concerning the Road at Horbury Bridge End particularly mentioned before.[290]

---

290 All West Riding bridges were the responsibility of the county. Magistrates in Quarter Sessions would make decisions and there were two West Riding bridge surveyors to oversee all work.

1761 Week 24 October 12th - 18th

Thursday 15th, Friday 16th

These Days at the Works, the Preceeding Part of the Week being confined by a Cold, sent my Clerk to enquire concerning the State of the Works after the great Flood that happened on Monday last, who bro[ugh]t an Acc[oun]t that nothing considerable had happened; the State whereof from my own Inspection is as follows.

At Wakefield Lock no damage, the Water had run over the Ground at the Lock head where the Banks had not been made up and likewise over where the Tunnell had been put in on each Side the Ground there not being made up before the Flood came, but the Damage upon the whole is very inconsiderable.

At Thorns Lock some small damage to the temporary Dam at the Tail but the Head Gates being hung and shut prevented the Water from making its way thro[ugh] the Lock but running over on both sides at the Lock Head where the Solid was not dug out and consequently the Banks not raised has taken away a few Loads of Loose Earth from the termination of the Banks on each side but made no breach in the Solid or produced any effect of consequence.

At Lupsit Dam the Water being ab[ou]t 3 Feet higher than the Dams End Walls, consequently 8 Feet over the Crown of the Dam it has taken the Earth from behind the Walls for 3 Feet deep; but without any Prejudice to the Masonry; it has also bro[ugh]t the Rubble Backing irregularly more forward upon the Crown of the Dam but without any prejudicial effect.

At the Cut in Horbury Mill Pasture no effect whatever. The Mason at work getting on the Caping there. At Washingstone Lock and Dam the Flood has taken no effect upon them and at Horbury Bridge Works the greatest effect is the taking a little of the Earth away that formed the Slope upon the new Bridge; ordered a small Rubble Arch to be turn'd therein and to raise the head of the Lock which is up to the Square 1 Course higher;

At the long Cut the Water having overtopp'd a part of the Bank had taken down ab[ou]t a Rood[291] in length which is the only Damage there, in Short all the Masonry is safe and secure and the Damage to the unfinish'd Part of the Cutts surprisingly small considering that the Flood was 8 Inches higher than any this seven year. I have now satisfy'd myself that these works when completed notwithstanding the rapidity of the River will not be subject to greater inconveniencies than others have suffer'd in a more quiet Situation.

Having some Business in Lincolnshire which I have postponed during the whole Summer in order to confine myself wholly to the Calder work I propose taking this opportunity of doing the same and propose to be back in a fortnight.

---

291 Rood applies to both length and area. In this sense a rood is about 5½ to 8 yards

1761 Week 25, 26 and 27 October 19th - November 8th

(Week 25 and 26, October 19th – November 1st in Lincolnshire)[292]

Week 27 November 2nd – 8th

Tuesday 3rd, Friday 6th, Saturday 7th
These Days upon the Works. On Tuesday Morning on my return from Lincolnshire inspected the Works to see what damage had been done by the Floods during my Absence; one of which on the Friday preceeding had been considerable, but found nothing stirr'd.

The State of the Works as follows.

The Hands in the Yard employ'd upon the Lock Gates for Horbury Mill Pasture and Washingstone & a Bridge over the Tail of Horbury Mill Lock

At Wakefield Lock a Breast work rais'd to prevent the higher Flood going over the Lock and the Banks made good to the same height.

At Thorn's Lock all the Gates hung and the Tail flue Walls completed, the Solid at the Lock Head dug out and the Banks made up there with the Solid also at Lupsit Bridge dug out the Banks made up and the Bridge made passable. The flue Walls of this Bridge completed; the Solid also clear'd to the Bottom between the Rock & lowest Part of the Cut, a part of the Rocky bottom being found too high was taken down to a proper Level in short all Impediments in this Cutt being removed by means of a Temporary Dam constructed with rabbited Piles, at the Head thereof the Dam was taken away and passage through the Cut opened on Friday;

and on Saturday the Tail Gates for Horbury Mill Pasture and also the tail Gates for Washingstone were navigated from the Work Yard to Horbury Mill Pasture Lock, At this Lock the Head Gate hung, and the Engine at work getting out the Water from the Chamber in order to hang the tail Gates; The Diggers at work in order to get out the Solid and make up the Banks at the lock Head and D[itt]o at the head of the Cut next the River. Carts employ'd in bringing Earth to make up the Ground next the Lock.

At Washingstone all there as p[er] last, the Engine got in order to take out the Water for hanging the Gates, the Head Gates of this Lock being also under Water.

At Horbury Bridge Lock getting on the Caping, the dry Arch under the Road being turn'd, hands at the Bridge getting up the Wing Walls; the Dam as p[er] last.

---

292 Smeaton was acting as consultant to John Grundy again, either on the Louth Navigation, (see 5th - 8th August 1760) or in connection with work on the River Witham. In 1761 John Grundy, John Smeaton and Langley Edwards submitted a report on a plan to construct a sluice on the river at Boston. (Ref: William Henry Wheeler *A History of the Fens of South Lincolnshire*, Boston and London 1897)

At the Long Cutt the Diggers advanced almost up to the Mill Bank Quarry, the late Rains have bro[ugh]t down a Thrust of Earth from the Bank that will require to be secured.

1761 Week 28 November 9th - 15th

Thursday 12th, Friday 13th, Saturday 14th
These Days attending the Works. Hands in the Yard upon the Plank Piling and the Lock Gates for Horbury Bridge.

At Thorns compleating the Rack & Wheel for drawing the Shuttles of the Gates; some Hands at Thorns Bridge and compleating the Wing Walls.

At Horbury Mill Pasture the Gates hung and Bridge over the tail made passable, the Solid at the head of the Lock dug out and the Banks made up. The Diggers employ'd in getting out the Solid at the head of the Cutt.

At Washingstone compleating the Caping and hanging the Head Gates, the tail Gates being got to Place, but the Chamber of the Lock cou'd not be emptyed since Wednesday's Flood.

At Horbury Bridge the Masonry of the Lock completed, some Hands employ'd upon the Parapet & Wing Walls of the Road Bridge.

At the long Cutt being advanced thro[ugh] the boggy Ground and the Banks made up near upon Mill Bank Quarry order'd the Hands to proceed to compleat the Junction between this and the lower Part of the Cutt.

1761 Week 29 November 16th – 22nd

Tuesday 17th, Wednesday 18th, Thursday 19th, Friday 20th
These Days attending the Works and the Meeting of the Commissioners at Halifax.

Hands in the Yard employ'd upon the Gates for Horbury Bridge Lock and upon a set of Tackle to be apply'd to the head or tail of a Lock for taking the Water off in order to hang Gates etc

At Horbury Mill Pasture Lock moving Earth in order to make good the Ground round the Lock. Diggers at work upon the solid at the Cut Head.

At Washingstone compleating the Foregates and making a Dam with rabbited Piles cross a Breach in the tail Dam in order to get out the Water from the Chamber to hang the tail Gates and the more effectually to get out the Core of earth from the Cut's tail; this being compleated on Wednesday the Engine began to work on Thursday and clear'd out the Water but on Friday Morning a Fresh again came down which filled the Lock Pit.

At Horbury Bridge everything as p[er] last the long Cutt advancing.

1761 Week 30 November 23th – 29th

Wednesday 25th, Thursday 26th
Wednesday Afternoon arrived at Wakefield and had the following acc[oun]t from Mr Gwyn[293] that the Fresh on Friday being scarcely run off, and the Engine got the Lock Pitt clear'd but a fresh downfall happening. Yesterday the Lock Pitt was again fill'd this day so that little cou'd be done towards the hanging of the tail Gates; but that the Shuttles, Racks and Wheels had been fix'd thereto and the other Works gone on with that were out of the Waters way.

On Thursday examined the Works in the Yard, the Head gates for Horbury Bridge compleated and proceeding with the tail D[itt]o.

At Thorns strengthening the Banks at the Lock head.

At Horbury Mill Pasture Lock the Engine at work in order to get out the tail Dam, the Hands employ'd in making good the Ground round the Lock Hills fitting up and levelling a Trench made for drainage, and making good the 12 foot Road agreed to be made for the People of Horbury between the 2 Parts of their Pasture otherwise disjoined by our Works; the Water turn'd upon the head Cutt Gate the Pott of which being set by a different hand proves somewhat leaky but may be rectify'd when the Water is taken off. Ordered a Breast work to be raised at this Lock Head like Wakefield to prevent the Waters overtopping the same in high Flood.

At Washingstone the Engine again at work.

At Horbury Bridge 2 Masons rectifying the head hollow Posts and Sells and letting in Cramps

1761 Week 31 November 30th - December 6th

Wednesday 2nd, Thursday 3rd, Friday 4th
Attending the Meeting of Commissioners at Halifax and upon the Works.

In the Yard at the Tail Gates of Horbury Bridge Lock.

At Horbury Mill Pasture Lock making the Breast Work and levelling the Bank for the communicating Road; the Tail Dam clear'd and cleaning away the Engine and Utensils and also the sunken Timber out of the Tail Cut.

---

293 This is the first mention of John Gwyn in the Journal. However the letters show that he was negotiating a salary to work as foreman under Nickalls in January 1760 and was not appointed to be superintendent of carpentry and smiths' work until 17th March 1762 at a salary of 50 guineas 'and his present salary to cease'. (Ref: WYJS/CA MIC2/2 and MIC2/4)

At Washingstone the Tail Gates compleated and the Tail of the Cut clear'd digging out the Solid at the Lock Head and making up the Banks; and for this Purpose a Dam of rabbited Piles being constructed at the Cutt head, the Cutt itself being somewhat too shallow ordered the same to be made deeper.

At Horbury Bridge the Masons letting in the Pots for the Head Gates and also the Collars & Cramps.

The long Cutt advanced to Mill Bank Quarry and set out Thornhill Lock in a Close call'd the New Ing near Thornhill Lodge.

1761 Weeks 32 and 33 December 7th – 20th

Week 32 December 7th – 13th No entries (see below)

Week 33 December 14th – 20th

Tuesday 15th, Wednesday 16th, Thursday 17th
These Days at Wakefield and attending the meeting of the Commissioners at Halifax, having been close confined the Week before in making the Plans of the Locks intended next year as per order of the Commissioners at the Meeting 2nd Inst. Called at Wakefield on Tuesday in my way to Halifax; but not having time to go upon the Works saw Mr Gwyn who reported that every [sic] was cleared except the Cut head at Wakefield, the Tail of Thorn's Cut and some of the old Timber from Horbury Mill Pasture Cutt, all which were then in hand that in pursuance of my order he had drove a Row of rabbited Piles cross the Head Cutt of the Pasture Lock and rectify'd the Pot as mentioned in my Journal of the 25th & 26th Ult. in consequence whereof the Head gate now shuts quite tight.

1761 Week 34 December 21st - 27th

Monday 21st, Tuesday 22nd
These Days upon the Works. Hands in the Yard employ'd in repairing the Engines and Pumps. The Head and Tail of Wakefield Cut cleared to 3 1/2 Feet deep. Thos Wood at work at the Tail of Thorns Cutt with the Rake clearing of that.

Some Hands employ'd in the tail Cutt of Horbury Mill Pasture in getting out the Sunken Timber, the Head Gate of this Lock shuts compleat. Boat and Hands at this Cut head drawing the rabitted Piles employ'd for taking off the water in order to get out the Core at the Cutts head and rectify the Gate.

At Washingstone Lock everything clear.

Saw Mr Plats who reported that Thornhill Lock Pitt from a hard Clay had turned to Gravel with Springs as usual.

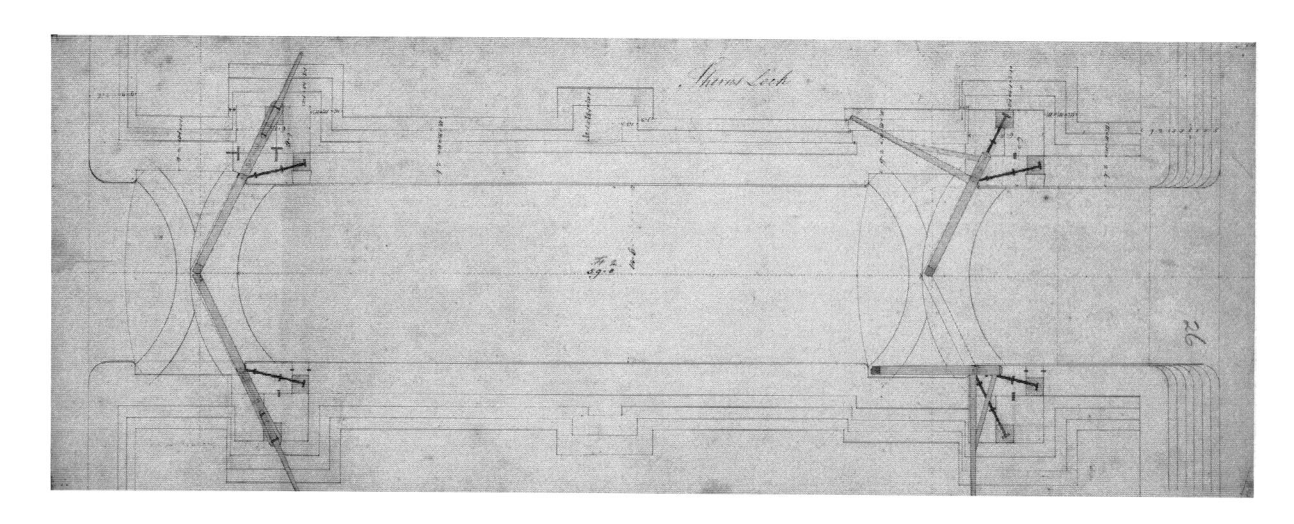

56. Plan of Thornes Lock showing the construction of the lock walls, the sills and apron, the hollow posts and the rotation of the lock gates.

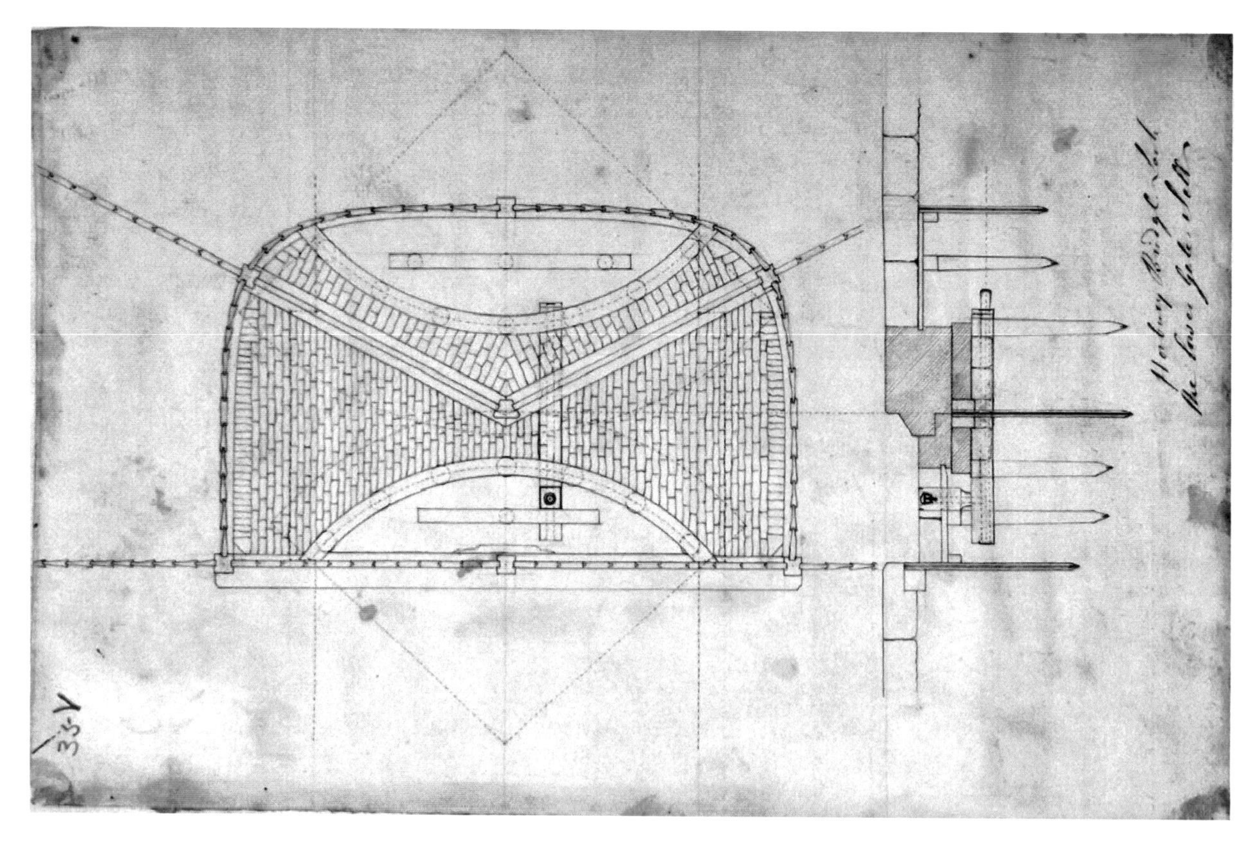

57. Plan of Horbury Bridge Lock, the lower gate sill, showing the piling.

1762

Week 1 March 29th - April 4th

Tuesday 30th
This Day arrived at Wakefield.

Wednesday 31st
This Day inspected the Works and attended the Meeting of Commissioners at Halifax.

Thursday April 1st Friday 2nd Saturday 3rd
These Days further examining the Works Shewing and explaining the same to Mr Scott[294] taking Levels etc

Found the State of the works as follows viz:

In the Work Yard the 3 large Engines ready for use and the Frame work of a new one in good forwardness and the Screw making.

At Wakefield Thorns and Lupsitt found everything in order as I left them, a Quantity of Rubble Stone having been thrown into a Pool formed by the Floods near the Skirt of Lupsitt Dam towards the South End and another at the Tail of the Cloughs, those Pools being so fill'd as p[er] order seem effectually secured, the Dam itself being perfectly entire.

The violence of the Water issuing from the Tail Clough of the high Lock in Horbury Mill Pasture seemed tending to form a Pool at the end of the Tail flue Walls, the same has therefore been secured (as p[er] order) by a few yards of wierring. The tail Cut of this Lock in which the floods made much Havock the preceeding Winter has stood the last perfectly well; but the rais'd Banks of the Head Cut of the same Lock have not fared quite so well, for in one of the Floods the Wind having set right down the Cut the Surge raised thereby has wash'd down some Parts of the face of the Bank above Soil which being thereby thinned will require to be strengthened before next Winter, and this damage tho[ugh] of small account I think is the most considerable that has happened during the Winter.

At Washingstone and Horbury Bridge everything is nearly as I left them except that the Road Way is now made good over Horbury Navigation Bridge. The Caping of the Pasture Lock Washingstone and Horbury Bridge Locks being laid on late in the Season the Pointing thereof has suffer'd somewhat by the Frosts and that of the Bridge Dam Cloughs by the floods all which are capable of easy remedy at the proper season

294 Mathias Scott, appointed surveyor and superintendent over the Masonry and Digging, 31st March 1762.

Found the long Cut advanced as far as the turn of the River just below Mr Banks Mill Dam[295] and the Men proceeding upwards by the Warren house within a Quarter of a Mile of the Cut Head all which has been done according to order.

On Wednesday 31 March the first Boat came up the River to Horbury Bridge loaded with Lime drawing 3 F[eet] 3 I[nches] Water, and on Thursday Ap[ri]l 1st the Piling of Thornhill Lock was begun.

The same Ev'ning went down with the Boat to Wakefield in order to examine the state of the Shoals, Cut heads and tails that had required dredging and which had been attempted by the Rake. Found some of them where the matter had been soft sufficiently cleared for present purposes but the Shoal at Horbury Cut head and another at the Tail of the Lock where the Dam had been made across the Cut being of an harder consistence, the Water was still there left too bare when in its ordinary state. And as almost all those Lock heads and tails together with the Shoals aforementioned will require further deepening, it appears necessary to have recourse (not only an acc[oun]t of the Present but of future exigencies of the same kind) to my first proposal of a Ballast Engine[296] by which those kind of operations may be performed to the desired degree.

<u>1762 Week 2 April 5th - 11th</u>

Monday 5th Tuesday 6th
These Days making a Design for a Ballast Engine settling Contracts &c......

Wednesday 7th to Saturday 10th
These Days inspecting the Works taking Levels setting out Ledger and Batty Mill Locks and Cutts sounding the Reach between Dewsbury low ford and Mr Greenwoods[297] Mill at Dewsbury, and also the Reach from Mirfield Low Mill to Ledger Mill in order to re-examine the Shoals in the same and found that the 2 lowest which consist chiefly of loose Bo[u]lders and extend but a few Yards. The Water being deep both above and below may be removed at a small expence and that the uppermost which is both the Shoalest and longest may be removed at a moderate expence or may wholly be avoided by extending the Cut about one furlong lower down the River than at first proposed which will make the same in the whole at about 2½ Furlongs in length.

Respecting the first mentioned Reach by Dewsbury Town found that deepening Watergate stream[298] for about 70 Yards and making it 2 Feet Deeper and raising the Dam at the Low

---

295 Mr Banks is shown as the owner of Dewsbury New Mill (also known as Sands Mill) in a key on Smeaton's map and the mill is marked on the map itself. (Ref: RS JS/6/9 Plan of the River Calder from Wakefield to Brooksmouth)

296 It is likely that the 'ballast engine' refers to a method of dredging using a succession of buckets to raise gravel, sand and mud from the bottom of the bed of the river or cut. See also 'Ballast Net' on the 15/16 April.

297 Mr Greenwood is shown as the owner of Dewsbury Mills in a key on Smeaton's map and the mill is marked on the map itself. (Ref: RS JS/6/9 Plan of the River Calder from Wakefield to Brooksmouth)

298 Watergate stream, just west of Dewsbury, marked as d on the shoals and streams panel of Smeaton's plan.

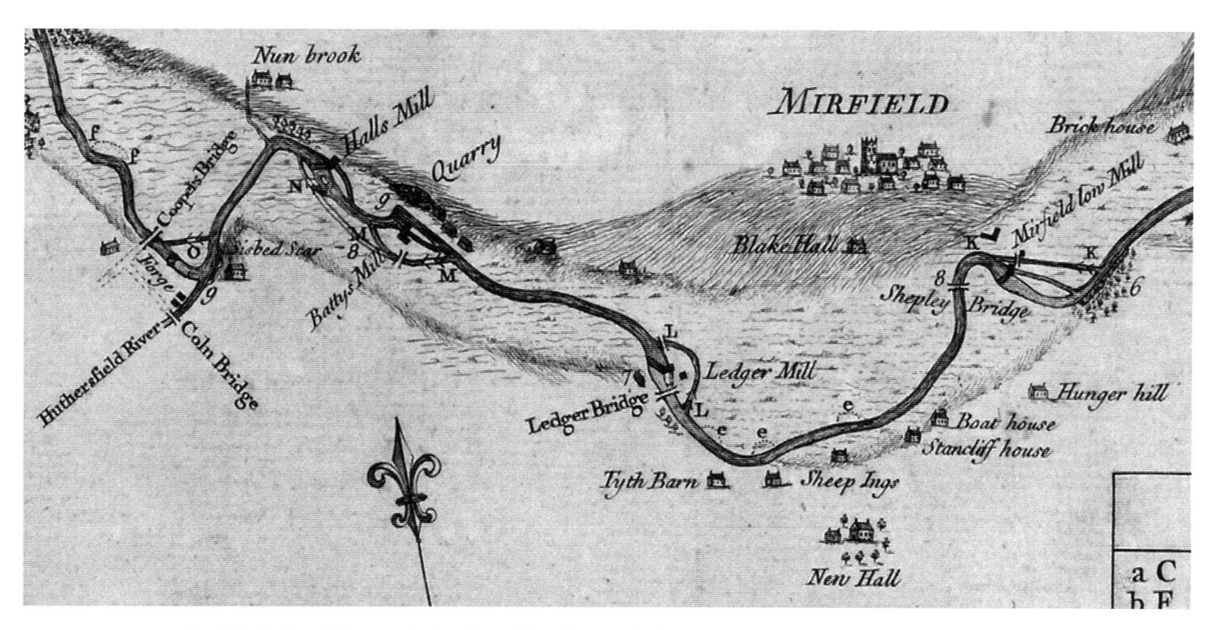

58. Mirfield and Cooper Bridge from John Smeaton's Plan of 1757/58

Ford so as to pen 18 Inches Water will render that Reach navigable from the Cut head Lock by Dewsbury Town to the tail of Dewsbury high Lock.

On Monday last the Diggers begun upon the lock Pit at Mirfield low Mill.

<u>**1762 Week 3 April 12th - 18th**</u>

Monday 12th
This Day making out the Journals of the two preceeding Weeks and went to Wakefield.

Tuesday 13th
This Day inspecting the Works the Piling Work at Thornhill Lock all done and the Workmen begun to Plank the Sell.

Wednesday 14th
This Day inspecting the Works and attending the Meeting of the Commissioners at Halifax

Thursday 15th, Friday 16th
These Days upon the Works and set out the Lock Pit at Ledger Mill according to the order of the Commissioners on Wednesday, but on reviewing the Ground and reconsidering the inconveniences that will happen in every direction thereof; in case the owner of the grounds shou'd happen to like the first intended tract thereof preferable to the latter, upon the whole I didn't think the preference worth disputing.

On Friday one of the Thresholds of the lower Gates of Thornhill Lock was laid down and fixed in Place, and the Masonry was begun. This Morning having discovered that the foundation of Figure 3 Lock was likely to turn out Rock summoned together all the Surveyors and the Undertakers upon the Place and as the foundation of Thornhill Lock was laid by the Plan designed to Figure 3 so the Plan intended for Thornhill becomes equally applicable to the Figure 3, therefore explained the same and method of proceeding thereupon in order to lay down a Stone low Sell &c and left orders that if any unforeseen difficulty shou'd arise immediately to let me know and if necessary wou'd return immediately.

This Ev'ning Rob[er]t Hartley's Boat[299] came up to Horbury Bridge loaded with 18 Doz[e]n Lime drawing 3 Feet Water (being either the 4th or 5th Trip) reported they had been stopp'd some time upon a Shoal in the old Navigation and had found the Water rather too short at Pasture Lock Tail occasioned by a Bed of Gravel thrown up by the lower Gate Cloughs therefore order'd Mr Gwyn as he had not hands enough to proceed with the Minion Mill and Ballast Engine both together, to defer the latter till the former shall be compleated and in the meantime to scrape away a little of the gravel aforesaid with the Ballast Net as heretofore and to put on a small Dam Board of ½ the Breadth of a Deal, the manner of fixing which being very simple I fully explained.

Having thus settled everything wanted with the Surveyors and having an opportunity of setting forward for London on Saturday proposed making use of the same.[300]

<u>1762 Weeks 4 -10 April 19th - June 6th</u> *During this period Smeaton was producing a report on the River Chelmer in Essex.*

<u>Weeks 4 - 9 April 19th to May 30th</u> *No entries*

<u>Week 10 May 31st - June 6th</u>

June Wednesday 2nd
This Day arrived at Wakefield and....

Thursday 3rd, Friday 4th, Saturday 5th
These Days inspecting the Works which were in the following state. viz.

In the Work Yard the new Engine and Screw was near upon ready for setting up, the Cogs of the stone Wheel want dressing off. Half of the Smith Shop removed to Dewsbury; some parts of the

---

299 Robert Hartley's boat was used on several occasions to assess the readiness for opening of a section.

300 This may be the period in 1762 when Smeaton attended Parliament on the Bills for Stockton Bridge and the Bridgewater Canal. (Ref: Skempton: John Smeaton Trevor Turner and A W Skempton. p 17)

301 Smeaton compared the quality of sample millstones to French stones in a letter of 1756. (Ref: Royal Society of the Arts, RSA/PR/GE/110/5/35)

Ballast Engine prepared; the Minion Mill made carried away and set up Horbury Bridge; the French Stones[301] arrived.

A Dam Board apply'd to the Crown of Lupsit Dam to carry the Vessels over the Shoal at the tail of Pasture Lock and tail of the Cutt. All the Ballast Engine can be got to work. The Banks in the Head Cut faced with Sods and made good.

Some Hands employ'd at Washingstone Dam to make good the temporary Dam in order to compleat the Principle Dam.

Horbury Bridge Lock as I left it viz. The Foregates hung the Stern Gates unhung, the flue Walls wanting and the tail Cut wants getting out all which wait for the working of an Engine to get out the Water.

At Horbury Bridge Dam the temporary Dams made in order to take off the Water, and the Masonry proceeding with 4 Bays, and clos'd them in on Saturday night. The Minion Mill at work, order'd all hands upon this Dam and Washingstone to remain till this is compleated.

The Rock Shoal below the Figure 3 near upon cleared. The Figure 3 Lock above low Water and compleating the tail flue Walls and everything under Water in order to let in the Calder.

At Thornhill Lock all the Masonry compleated except the Head flue Walls. The long Cut advanced to the Cut Head leaving Solids in proper places and the Lock Pit there dug ready for the Piling.

Dewsbury Mill Lock or as it is more frequently called by the Workmen Watergate Lock, the Pitt sinking and Cut advancing.

At Mirfield low Mill Lock the Walls almost up to low Water mark. The Barrow Lime arrived at Horbury Bridge in one of the largest Gainsboro' Keels that can possibly come up these Rivers. This Keelman had with much Industry been persuaded that he cou'd not pass the Bridges and yet upon tryal found above 3 Feet to spare............

1762 Week 11 June 7th - 13th

Thursday 10th Friday 11th Saturday 12th
These Days attending the Works upon the Calder.

Washingstone Dam as p[er] last.

At Horbury Bridge the Carpenters and Labourers piling the Foundations of the South end Wall and getting 3 Bays contiguous thereto ready for the Masons, having cleared the Ground for an additional Bay as ordered last Week.

At the Figure 3 Lock the Walls almost up to the Square. Thornhill Lock as p[er] last. The Bridge building for passage to Thornhill Lodge. The Head Sell of Cut head Lock pitched and begun to Pile.

Watergate Lock the Pitt sinking and Cut advancing.

At Mirfield Low Mill Lock Walls above Water, Stern flue Walls Building and Bottom setting,

Ledger Mill Lock Pit begun.

1762 Week 12 June 14th - 20th

June Wednesday 16th Thursday 17th
These Days at the union Club and attending the Annual Meeting of Commissioners at Halifax

Friday 18th Saturday 19th
These Days attending the Works upon the Calder

Horbury Bridge Dam the South End Wall near Dams height and the Masons proceeding with 3 Bays contiguous; the Carpenters and Labourers proceeding with the remaining 7.

Figure of 3 Lock up to the square.

Thornhill Bridge D[itt]o. At Cut head the upper Sell near compleated the lower D[itt]o pitch'd and piling; The Lock Pit and Cut at Watergate going on as p[er] last.

At Mirfield low Mill the Bottom compleated and the Water let in, the Walls advancing.

Ledger Lock Pit proceeding as p[er] last.

1762 Week 13 June 21st - 27th

June Wednesday 23rd Thursday 24th Friday 25th
These Days upon the Works.

The Carpenters compleating the last 4 Bays at Horbury Bridge Dam and the Masons proceeding with the Rest.

Some hands employ'd in cutting away the slip of Earth in the long Cut and piling a foundation for a Wall, but finding a Rock 3 Feet under the bottom judged it adviseable to get down the Wall upon the Rock. Mr Walker's Bridge[302] on the Lock Cut begun.

---

302 The identity of Mr Walker and the position of his bridge have not been identified, but he may be the Thomas Walker of Headfield whose burial in June 1768 is recorded in Thornhill parish register.

Some Hands employ'd in banking Noel's Pond near the Warren house[303] and making a Road by the same. The lower Sell of Cut head Lock nearly compleated.

Watergate Lock Pit down at the Calder Surface and getting up the Engine.
At Mirfield Low Mill the upper Sell set and the Walls advancing.

Ledger Lock Pit advancing.

On Wednesday last the Works were inspected by Mr Caygill & Mr Stansfeld.

1762 Week 14 June 28th - July 4th

July Thursday 1st Friday 2nd
These Days attending the Works upon the Calder.

Labourers employ'd on digging out the foundation for the North end Wall of Washingstone Dam and making temporary Dams and driving Piles and the Carpenters employ'd in fixing tye Beams etc for the 3 Bays left uncompleated last year, which with an additional one now proposed makes 6 which want compleating.

At Horbury Bridge on Tuesday Ev'ning the Body of the Dam near a Close expected to be compleated by Saturday noon.

The Slip of Earth in the long Cut as p[er] last the hands being otherways employed.
The carpenters putting on Thornhill Bridge, the Walling of Mr Walker's considerably advanced.

Hands proceeding with the Road at the Warren house and banking of Noel Pond.

Cut Head Lock the Timber Work of the Foundation compleated, but the Masonry not begun, all Hands being employ'd at Horbury Bridge Dam and Walker's Bridge.

At Watergate the Engine fix'd and proceeding with the Bottom of the Pit.

At Mirfield low Mill the Lock Walls at High Water.

Ledger Lock Pit proceeding as per last......

---

303 Warren House is shown on Smeaton's plan, but Noel's Pond has not been identified. However an abstract of ownership within the Savile collection at Kirklees Archives (Ref: DD/S/1/217) notes that between 1679 and 1757 Nowell Farm was in the possession of the Nowell family but owned in 1757 by Richard Wilcock. Both this and Bottoms Farm which was owned by William Wilcock were listed together. On the 1849 OS map Bottoms Farm is shown west of Warren House.

1762 Week 15 July 5th - 11th

Wednesday 7th Thursday 8th
Attending a propos'd Meeting of Commissioners at Halifax.

Friday 9th Saturday 10th
These Days attending the works upon the Calder.

The Carpenters and Labourers at work in Piling the Foundation and fixing the works of Washingstone Dam. The Rain which fell on Wednesday afternoon Thursday & Friday Morning interrupted not only this but all the other Works by driving the Workmen there from the Diggers not excepted, but the Calder was scarce sensibly affected thereby.

Horbury Bridge Dam clos'd in last Saturday and the South End Wall advanced 3 Course above Dams height.

The Caping of Thornhill Lock compleated;

The Mason finishing the Wing Walls of Thornhill Bridge.

Mr Walker's Bridge ready for the Carpentry.

All at Cut Head in the same state as p[er] last.

At Watergate the Lock Pit near finished but full of water while the Hands were beat off by the rain.

At Mirfield low Mill little done this Week. The hands having been employ'd in preparing Stuff for Ledger Mill Lock.

The new Engine set up at Ledger Lock which performs exceeding well and the Lock Pit almost down.

1762 Week 16 July 12th - 18th

Wednesday 14th Thursday 15th Friday 16th
These Days attending the Works upon the Calder.

At Washingstone Dam the Masons begun upon the same and the Carpentry far advanced but hind[e]red by the frequent Showers.

At Horbury Bridge Dam all the Masonry compleated except the Caping of the South End Wall.

Some Hands Piling the Slip of Earth in the long Cut.

At Cut Head Lock the Foundation got above Water and getting in the Bottom.

At Watergate Lock the Carpenters piling the Bottom but as this Locks Bottom turns out not materially different from the Rest and the Lower Gate Sells employ'd this year seem to answer extreamly well, it is judged that this Locks Sell by extending the Apron would fill the Stern Recess together with an additional Row of Feather edged Piles will answer equally well as the rest and be considerable saving to the work. This step is also the more necessary as the Carpenters have been under difficulties in providing materials for a Bottom wholly Timber, and in consequence will produce a considerable delay to the Masonry.

At Mirfield low Mill everything above high Water but nothing done here this Week, the Hands being employ'd at Washingstone.

At Ledger Mill the Lock Pit near down but not compleated.

The side of the Tail Cut at Washingstone Lock being of so soft matter, the action of the Cloughs in the tail Gates of the Lock has bro[ugh]t them down and landed up the tail of the Cut so as to prevent the Passage of the Vessels unless Horbury Mill Dam is quite full therefore ordered Mr Gwyn to wharf up the same with a proper wear and so far to open the Passage as shall be necessary till the Ballast Engine is completed.

1762 Week 17 July 19th - 25th

Thursday 22nd Friday 23rd
These Days attending the Works upon the Calder.

At Washingstone Dam the Carpenters work compleated and the Masonry advanced so far as to close tomorrow morning. The North End Wall raised above Dams height.

At Horbury Bridge Lock setting up the Engine for compleating the same. Horbury Bridge Dam as p[er] last.

Masons compleating the Caping of Figure 3 Lock & building the Wall at the Slip of Earth in the long Cut.

Carpenters getting on the Timber work of Mr Walkers Bridge.

Cut head Lock considerably above water.

The Piling for fencing off the Water at Watergate stream begun. At Watergate Lock the Carpenters laying down the lower gate Sell and Apron and proceeding with the Piling.

At Mirfield low Mill Lock getting on the Caping.

At Ledger Mill Lock the Lock Pit down & the Piling begun.

The Lock Pitt at Batty's Mill begun. Left orders with Messrs Gwyn and Scott for clearing and wearing Washingstone Lock tail as mentioned in my last and also to take the opportunity of doing the same at Pasture Cut Head etc

1762 Week 18 July 26th - August 1st

Saturday 31st
This Day over the works.

The Masonry of Washingstone Dam completed except part of the tail flue Wall and part of the Backing of the North end Wall which was built too high by mistake of orders. Washingstone Lock tail cleared and weared[304] but the Jutty not quite compleated.

The Head of Pasture Cut cleared and also the Tail Cut, while the Water of Lupsit Dam was drawn down for building an Horse Bridge at Dorcar[305] Lane End for the Bridle Road which was also compleated and this Day the Hands proceeding to clear the head of Wakefield Cut.

At Horbury Bridge Lock the Water got out, the Carpenters & Masons proceeding with the hanging of the Tail Gates, and the Digger getting out the solid at the locks tail, at the Dam Part of the Caping got on to the south End Wall.

The Masonry of Fig 3 Lock compleated as also the Wall at the Slip of Earth.

Cut Head Lock the Chamber Walls at their height, the Breast work at the Head advancing.

At Watergate stream 12 Bays of the fence Piling compleated and the rest advancing. At Watergate Lock the Thresholds of the low Gates and side string pieces yet to lay on acc[oun]t of the Water not being kept sufficiently out of the Lock which was owing to the Engine being out of order.

At Mirfield Mill Lock all the Masonry compleated.

At Ledger Mill laying down the low Sell and proceeding with the Piling.

At Batty's Mill the Lock pit below Water and proceeding with one of the hand Screws.

304 provided with a weir.

305 Dorcar – Durkar, south of Wakefield.

1762 Week 19 August 2nd - 8th

Thursday 5th Friday 6th Saturday 7th
These Days upon the Works all below Horbury Bridge clear and the Masonry completed.

At Horbury Bridge Lock Carpenters, Masons and Diggers employed in hanging the Gates, building flue Wall and rectifying Sell and hollow Posts, letting in Gate Collars Cramps etc and digging out the Lock tail, which work was interrupted by the rain on Friday and a fresh which came down on Saturday Morning.

Mr Wilcock's Bridge near the head of the long Cut Piled and the Masons begun upon the same.

The Lock at Cut head not completed, the Stone for that Purpose being blocked up in the Quarry. On Saturday set out the Cut head Dam.

The fence Piling at Watergate Shoal advanced to 20 Bays and proceeding with the Rest. At Watergate Lock the Chamfer'd Course got on the length of the Chamber, but interrupted as well as all the rest by the Rain.

At Ledger Lock the Carpenters at work upon the Foundation.

At Batty's Mill the Diggers proceeding with the Lock Pit with the hand Screw under Calders Level.

1762 Week 20 August 9th - 15th

Thursday 12th Friday 13th Saturday 14th
Attending the Works.

At Horbury Bridge Lock the Gates compleated and everything below water except the Digging of the Lock tail which had been completed this Week, but for the rain of Friday and Flood on Saturday which rose about 2 Feet upon the Crown of the Bridge Dam and stopp'd all the works.

At Figure 3 getting ready the Gates for hanging.

Mr Wilcock's Bridge near Cut head building, Cut head Lock as p[er] last.

Carpenters at work upon the Dam at Cut head; and upon the fence at Watergate stream being advanced 23 Bays, Watergate Lock the Masonry advanced below Calders Level,

At Ledger Mill the Masonry begun and part of the chamfer'd Course on.

The Digging of Mirfield low Mill Cut begun, and of Batty Mill Lock advancing as per last.

1762 Week 21 Aug 16th - 22th

Wednesday 18th
This Day at Halifax to attend a Meeting of Commissioners.

Thursday 17th Friday 18th [should be 19th and 20th]
These Days attending the Calder Works

Everything completed at Horbury Lock except the Core of the Tail below water, the Dam there having, by weakening the same too suddenly, broke in: ordered the Pumping to be knocked off and the same to be clear'd by dredging when the Water can be drawn off; and in the meantime a passage to be cleared for Boats to pass when the Dam below is full.

At Thornhill Lock the Tail Bridge laid over and making up the Road, putting together the tailgates and hanging the Head Gates of D[itt]o. Diggers getting out the solid at the Tail.

Mr Wilcocks Bridge the Masonry nearly got up to the Square.

Cut head Lock as p[er] last. Carpenters at work upon the Dam;

The Fence at Watergate Stream wanting 2 Bays of closing at the Head, Watergate Lock the Masonry advanced near low water mark.

At Ledger Lock the Masonry advancing below Calders Level.

On Friday waited on Mr Richardson[306] with orders from Messrs Caygill and Stansfeld relative to the Purchase of a Piece of Ground for a Warehouse at Dewsbury; & bid him 5d p[er] Yard for what we shou'd want, to which he promised to return an answer in a fortnight.

1762 Week 22 Aug 23nd - 29th

The Calder Works being all in due course last Week attended a Survey of the Foss Dyke Navigation and land adjoining from the Trent to Lincoln.[307]

---

306 William Westbrooke Richardson (1725-1771) Lord of the Manor of Dewsbury. A letter from Smeaton, 14th Aug 1762, to Mr Simpson suggests that stone had been removed from a quarry without the necessary permission from him as Lord of the Manor. The proposed warehouse was also discussed. (Ref: WYJS/CA MIC2/40)

307 Smeaton was again working with John Grundy on improvements to the Fossdyke Navigation (Ref: Report of John Smeaton and John Grundy, Engineers, concerning the Practicability of improving the Fossdyke navigation, and draining the Land laying thereupon, from a view and levels taken in August 1762, Cambridge University Press 2014)

1762 Week 23 August 30th - September 5th

Tuesday 31st Wedn Sepr 1st Thursday 2nd Friday 3rd
These Days upon the Calder and at Hallifax to attend the Turnpike Meeting[308], Union Club and Navigation Meeting.

This day 3rd [Friday] upon the Works of the Calder.

Everything clear at Horbury Bridge for the Passage of Vessels with a full Dam.

The Tail and Core at Figure [3] got out and Passage cleared into the Lock the Gates of this Lock all hung and making the Bridge for the Tail.

At Thornhill Lock the Gates all hung and Bridge passable. The Diggers at work upon the Tail Solid. Pipes laid under the Bank at Thornhill Bridge for conveying the Water of a Spring underneath the Cut for the use of the Lodge as before the Cut was made.

The Diggers taking out the Solid and making up the Road over Mr Walkers Bridge.

Mr Wilcocks Bridge the Timber work completed and making up the Road.

At Cut Head Lock the Masons laying on the Caping and Carpenters making the Gates.

At Watergate stream the Fence piling closed at the head and very water tight. In attempting to close on Saturday Se'nnight upon the rise of the River the Ground at Bottom was blown away before the piles coud be got down and the Freshes have interrupted a second attempt till this Week. At Watergate Lock the Masonry above Water but the Floor not set.

At Ledger Mill the Walls above Water, setting the floor and piling for the tail flue Walls.

At Batty's Mill emptying the Lock Pitt having been filled with the Freshes last Week in order to bottom the same.

---

308 This may have been a site meeting along the Turnpike as an order was made and a contract signed by two of the Trustees, Richard and Benjamin Cooke, probably on 2nd Sept. Moses Webster was to temporarily repair two sections of the road, from the Top of Halifax Bank to Shibden Hall Bridge and from Belly Bridge to Birtby Lane before October. The repairs involved laying 150 and 100 cartloads of stone respectively along these sections. Smeaton may have consulted on this work. The Trustees involved on this Halifax stretch were many of the same names as the Halifax Commissioners: Caygill, Stansfeld, Dr Jackson etc. (Ref: WYJS/WA, RT 105, Wakefield and Halifax Turnpike Road 1757 – 1776, Halifax District.)

1762 Week 24 September 6th - 12th

Wednesday 8th Thursday 9th Saturday 11th
These Days upon the Works.

On Thursday last being an heavy rain came down a large and sudden flood in the night between Monday & Tuesday which overtopping the natural Surface of the Ground above the Dam at Horbury Bridge on the South Side of the River and the Swarth[309] having been broken by the tracking of Carts, near the end of the Dam the force of the Current in pursuing those tracks worked itself a new Passage through the solid ground leaving the Dams End Wall with every part of the Dam perfectly sound and entire.

On Wednesday Morning being then acquainted with this Accident, went over to Horbury Bridge and found the whole Current of the River flowing through this Passage being in width from the upper end of the Dams end Wall to the nearest Part of the Grass in the West side of the breach 41 feet, having enlarg'd itself to the South as far as the little Beck it had taken away in the whole 3 Roods of Land, being common pasture Ground belonging the Town of Horbury. Immediately ordered a number of hands to be employed in getting Rubble and the Flatts[310] in bringing up the same to the Breach in order first to secure the Ground and next to block up the Breach therewith in the same manner as the Dam at Washingstone was secured the last year but not having much Rubble beforehand and one of the Flats having been made leaky by false Loading in getting out the Figure 3 Lock tail only one Flatt coud be employ'd till Saturday night and that with much embarrassment, as the Flood Water waining off thro' the Breach left so little water in the Cut above the Bridge lock that the Flatt was obliged to come up with half Loading, ordered therefore both Flatts to come up with a whole loading to Bridge Lock tail there to lighten till they cou'd pass the Cut and Carts to bring up the Rest on Sunday till the Breach could be so far stopped as to make Navigation for the Flatt when full loaded.

The Situation of the other works as follows.

At Figure 3 the tail Bridge compleated. Diggers getting out the head solid and backing up the Bridge, taking out the Solid at Thornhill Bridge, Backing up Mr Wilcocks Bridge and Diggers getting out the tail Solid, making the Gates at Cuthead, closing in the tail at Watergate Stream. At Watergate Lock getting on the fore Sell.

At Ledger Mill getting out the water in order to Bottom the Flue Walls & at Batty's Mill getting out the Water in order to bottom the Lock Pit. N.B. the late Flood overtopp'd the Solid at Cut head at Cut head [sic] and at the head of Watergate Lock so as in part to fill the Long Cutt and wholly the latter but without material Damage.

---

309 swarth (or sward) - the grassy surface of land.

310 flatts - punts or boats with a shallow draft.

1762 Week 25 September 13th - 19th

Monday 13th Thursday 16th Friday 17th
These Days attending the Works.

Some Rain having fallen in the West on Saturday Night produced a considerable Fresh on Sunday, Yet the Ground being tolerably secured at the Breach, tho[ugh] a small addition of Land was lost yet it was in no material Place, and the Flatts and Carts so far supply'd the Breach with Stone that on Monday the Navigation was restored for the Flatts: and on Tuesday one of Mr Birt's Keels[311] came up with Corn for Mr Wilcock and having delivered ⅓ of his loading at Horbury Bridge proceeded with the rest to Figure 3.

On Thursday the Water was penn'd Dams height, by the Catch Dam so as fully to restore the Navigation to Fig 3 and on Friday the Banks of the Breach was pretty well lined with Rubble so as to be in a likely way to secure the same from further Attacks of the Freshes. Some Rain falling on Wednesday Evening produced a small Fresh on Thursday but without any detriment.

This Week building a Wharf Wall at Mill Bank Quarry, getting out the Solid at Mr Wilcocks Bridge and clearing the Cut. Carpenters hanging the Gates at Cut head.

The fence at Watergate Stream being securely shut in and the hand Screw fixed for getting out the water. On Thursday noon as soon as the fresh was wained from running over the Heads of the Piles they begun to work and on Friday Morning the Diggers begun to take out the matter. At Watergate Lock the Foresell completed & work advancing above Water.

At Ledger Lock Bottom and flue Walls all compleated below Water and the Works above advancing.

At Batty Mill the Lock Pit being bottomed, on Friday the Carpenters began to pile the foundation.

1762 Week 26 September 20th - 26th

Monday 20th Tuesday 21st Thursday 23rd Friday 24th
These Days upon the Works this Week at the Breach at Horbury Bridge.

The Catch Dam being compleated was to pen the Water over the Shain Dam[312], and the Sides being thoroughly lined with Rubble in order to secure the Ground begun to fill up the deep Part

311 In 1758 Peter Birt and and Sir Henry Ibbetson acquired a 14-year lease on the tolls of the Aire and Calder Navigation for £6,000. Birt owned a large number of craft working on the navigation. In 1772 improvements were long overdue and Smeaton was called in to make proposals which were implemented by Jessop from 1775. (Ref: Peter L Smith , The Aire and Calder Navigation, Wakefield, 1987.)

312 meaning not identified, probably 'chain dam'.

of the Gullett where the main step is intended to be made. Several Keels with Lime the latter end of last Week and this Week delivered Cargoes at the Figure 3.

Some Rain happening on Sunday Night the Calder on Monday Forenoon trickled over the Fence at Watergate Stream however not so much but that the Screws kept it at an under while it subsided without producing any hindrance. So that on Tuesday Night the Matter was got out everything cleared and the Water let in. On Wednesday began to draw the fence Piles at the Tail. On Thursday and Friday hands emply'd in distributing the Matter taken out and in wharfing up and securing the same from going in again; and further to prevent which a Tunnel was begun for conveying the Water which runs down from Daw Green[313] in sudden Rains with great rapidity in the Calder at Watergate into the same by a different Passage.

On Monday the Carpenters begun again on Cutt head Dam and on Wednesday the Diggers began again upon Cutt head Solid. Thursday Afternoon produced rain which hindered the Work that Day, and on Friday the Fresh prevented the Carpenters from proceeding with the Dam.

Watergate Lock advancing above Water.

At Ledger Mill the Upper Sell got on and work advancing above Water; the Fresh rose in this Lock 3½ feet perpendicular.

At Batty's Mill the Carpenters going on with the Piling but the Fresh on Fryday Morning fill'd the Lock Pitt which stopp'd their Proceeding for the Present.

<u>1762 Week 27 September 27th - October 3rd</u>

Friday 1st Saturday 2nd
These Days upon the Works.

Wednesday and Thursday having proved rainy a considerable Flood came down on Friday, but the Breach being pretty well secured produced no alteration there, nor indeed to any part of the Works except the filling of them with Water.

The Solid at the long Cut head being cleared on Wednesday night, being Michaelmas Day, the Water was let into the Cut, but the core cou'd not be got out by the Diggers on account of large roots of Trees that had formerly grown on the Bank Side and large Rubble Stones that had been thrown in for preserving the Banks. On filling the Cut all the Banks proved as tight as cou'd be expected except at a place where a Tunnell had been laid for supplying Widow Walkers[314] Cattle

313 Daw Green was the western part of Dewsbury near Crow Nest.
314 Widow Walker has not been identified.

with Water which either not having been sufficiently ramm'd or the Ground having shrunk by the former dryness of the Season a Blow appeared but by the Assiduity of Mr Gwyn the Same was stopp'd on Thursday and the Flood being kept out of the Cut by the Gates at Cut head no damage ensued therein by the Flood on Friday.

all the Gates proved sufficiently staunch except one of the Tail Gates at Figure 3, the former by being too hard bound in the Collar and twining Post, the latter in being too hard upon the Chamfer so that they cou'd not meet the Sell both which were ordered to be rectify'd the first Opportunity. It appearing that some mischief might ensue to the Banks of the Cut and Ground adjoining from evil minded People's drawing the Cloughs of the Cut head Gates in time of Floods; bye Washes or tumbling Bays[315] were ordered near Figure of 3 & Thornhill Locks to take off such surplus water,

the Gage Piles at Cut head dam were all drove the string Pieces fix'd and the Plank piling drove except 8 Bays, those vacant Bays afforded a sufficient easement to the Waters Passage during the flood so as to prevent its blowing away the Bottom.

In the Interval of the floods since last Week Mr Gwyn had drawn the Piles at the tail of the Fence at Watergate Stream and had endeavour'd to do the same at the head, but the Piles were drove so deep and grown so fast together that the same means proved ineffectual so that they stood the Brunt of the present Flood notwithstanding all the Braces and Shoars were taken away; so that this penn'd the Water to a considerable height but without any ill effect except taking away some Matter from Bank where the Piles joined the Land but this was stopp'd by throwing in some Rubble.

Watergate Lock advancing above Water, Ledger Lock above water as p[er] last;

At Batty's Mill the Leakage being very great the Water was scarcely got out when the Flood on Friday came on and again filled the same:

having occasion to be abroad next Week, everything was settled with the Officers accordingly

1762 Week 28 October 4th - 10th No entries.

1762 Week 29 October 11th - 17th

Wednesday 13th Thursday 14th
These Days attending the Works which were as follows,

---

315 Bye washes or tumbling bays were used on many navigations and canals to prevent the level of the water above the lock becoming too high.

Something done at Horbury Bridge Cut Banks which had been ordered to be strengthened. The Men employ'd about the Breach had proceeded to fill up a deep hole where the main Stop is intended to further strengthen the lining of the Banks against the Land and to raise a Bank of rubble upon the Top of the Natural Bank to prevent the high Floods from taking the same liberty with fresh Ground near the Dam as before;

Something done towards the Bye Wash at Figure 3. The Soakage of the Cut was very inconsiderable but some Back Drains appeared necessary which were ordered.

The Core at Cutt head is in a great measure got out sufficiently for present navigation, and the Roots cut up; on Tuesday the Cut head Dam was clos'd in but the Water having been penned by a Catch Dam of rubble for the Apron made to facilitate the closing thereof produced the proper navigable Water, so that Rob[er]t Hartley's Vessel full loaded with Lime went through Dewsbury Reach and d'[elivere]d her Cargo at Watergate Lock, on Sunday the 10 Inst being Old Michaelmas Day;

The fence at the head of Watergate Stream being removed and cleared there is as good Navigation now where that Shoal was as any Part of the River. The first Vessel that came up the long Cut was on Thursday the 7th but the Core of the Cut head which not being then sufficiently cleared and the Dam not clos'd in, there was not water over the same so delivered her Cargo in Cut head Lock;

The Cut being now filled to its proper height it appears that there will never be less than 3F 6I over the highest Threshold in this Cutt.

At Watergate Lock having got out the water, Hands proceeding to set the Floor and build the tail Flue Walls.

Diggers proceeding with the Cut.

Ledger Mill Lock as p[er] last.

Battys Mill Lock the Undertaker having fixed 2 Engines thereupon and thereby conquered the Water the Piling was finished. The Chamfer'd Course got on and the Floor set and proceeding with the Piling for the Flue Walls.

<u>1762 Week 30 October 18th - 24th</u>

Wednesday 20th Thursday 21st Friday 22nd Saturday 23rd
These Days attending the Calder Works.

At Horbury Bridge Dam the Ground being well secured, Hands proceeding to raise the Body of the Main stop, the Water being there deep and requiring a considerable Base takes a great Quantity of Rubble to raise the same above Water.

The Bye Wash at Figure 3 compleated, the Piling of Cut head Dam and all the Carpenters work completed to its proper height but wants a quantity of rubble to extend the Apron and further secure the South end; on Wednesday night hard Rain which filled the River Bank full on Thursday but no damage here or elsewhere.

At watergate Lock, having set the Floor, built the Stern flue Walls and compleated everything below Water, the River was let in and the Hands proceeding with the Walls which were raised above high Water mark. Masons at Work upon the Middle Bridge next Mr Greenwoods, and the Diggers proceeding with the Cut.

At Ledger Lock everything as p[er] last only the Diggers proceeding with the Cut.

At Batty's Mill all the Masonry brought near low water mark, but the Flood on Thursday filling the Lock Pit put the Hands off work but however the Water being in due time let in, prevents the Blowing of the tail Dam and on Friday the Hands got to work again.

<u>1762 Week 31 October 25th - 31st</u>

Thursday 28th Friday 29th Saturday 30th
These Days upon Calder works.

At Horbury Bridge raising and strengthening the Banks of the Cutt, Masons compleating the Battlements of the Cut Bridge, Labourers throwing in Rubble to the Dam Skirt and the Breach.

Upon the long Cutt the Hands employ'd in cutting out the Bye Wash at Thornhill Lock head, making up and strengthening the Banks, stopping the Oozing of Water, making Back Drains &c.

At Cut head Dam throwing in Rubble.

At Watergate Lock one Side up to the square but the Masonry here as elsewhere much interrupted by rains and Frosts. Diggers proceeding with the Cut.

At Ledger Mill Diggers proceeding with the Cutt.

At Batty's Mill all the Work above low water except one course of Arch Stones for the Forebay and one of the tail flue Walls which wants raising above a Foot which had been retarded by the Frosty Mornings. On Friday Morning snow'd very hard which did interrupt everything for that Day[316].

---

316 The heavy snowfall of 29th October, was followed by a particularly cold December and January. April to July 1762 had been a prolonged spell of warm fine weather. (Ref: http://booty.org.uk/booty.weather/climate/1750_1799.htm)

1762 Week 32 November 1st - 7th

Wednesday 3rd Thursday 4th Friday 5th Saturday 6th
These Days attending the Calder Works and the general meeting of the Commissioners at Halifax.

At Horbury Bridge Masons compleating the Battlements of the Cut Bridge. Labourers throwing in Rubble to the Dam Skirt & Breach.

Some Diggers employ'd in putting things to rights upon the long Cutt.

At Watergate Lock the Caping compleated on Saturday and proceeding with the head flue Walls; Diggers proceeding with the Cut.

At Ledger Mill also the Diggers proceeding with the Cut.

At Batty's Mill the Walls above low Water; the Water let in and the Engine taken out; but the Season for building being very bad the Walls were ordered by the Commissioners to be brought to a square and discontinued for the Winter.

1762 Week 33 November 8th - 14th

Friday 12th Saturday 13th
These Days upon the Calder Works.

At Horbury Bridge Dam the Labourers proceeding to fill up the Breach but little done this week on acc[oun]t of a considerable Flood that came down on Monday in consequence of which the Water continued high till Thursday.

The long Cut being emptied of Water during the Flood Mr Platts took that opportunity of Stopping the Leak. Some Hands employ'd in levelling the Cutt head Lock hill and making up the Banks to the Wing Walls at the head of that Lock.

At Watergate Lock all the Masonry compleated except one of the Head Wing Walls which the Floods had prevented the Masons proceeding with who were therefore employ'd upon the Bridge near Mr Greenwoods.

Diggers at Work upon the Cutt and the same at Ledger Cutt.

Battys Lock as p[er] last the Floods having prevented any means of proceeding therewith.

1762 Week 34 November 15th - 21st

Thursday 18th Friday 19th Saturday 20th
These Days attending the Meeting of Commissioners at Halifax and upon the Works.

At Horbury Bridge Dam the Labourers having proceeded to fill up and close in the Breach so that the main Stop takes the Pen of Water and throws the same over the Dam which being now in a state of security the Labourers were ordered to proceed with filling up the Pools at the Skirt of the Dam.

At Watergate all the Masonry completed, except one of the Wing Walls which had been hindered by the Frosts & Snows.

Hands at Work upon Mr Greenwoods Bridge. Diggers at the Cut and nearly completed up to the Solid.

At Ledger Cutt the Diggers proceeding therewith, it being open Weather on Friday the Masons at Work in levelling up the Walls of Battys Mill Lock.

1762 Week 35 November 22nd – 28th No entries

1762 Week 36 November 29th - December 5th

Thursday 2nd Friday 3rd Saturday 4th
These Days attending the Meeting of Commissioners at Halifax and upon the Works.

At Horbury Bridge Dam the Breach being as before the Labourers had made some progress in filling up the Pools at the Dam's Skirt but the Freshes having left a Bank of Sand within the Tail Cut of the Bridge Lock ordered the Labourers to proceed immediately to raise the Stone Jetty with Rubble in order to prevent the said effect. Also ordered Mr Platt to make up the Bank between the Bridge and Lock on the south side of the Cut, so as to form a Road between the Bridge and Lock Hill which is a Place where Goods may be delivered.

At Figure 3 Carpenters making a Carriage Bridge over the Waste Water Drain in order to make a communication with the Lands & Bridle Road.

At Thornhill Lock the Bye Wash compleated and making the Drain therefrom, a part of which being to be arched over for making a necessary communication they were digging the Excavation for the same.

At Watergate Lock the Flue Walls compleated and making some dry rubble Walls at the Tail for shoaring up the made Earth; the Tail Bridge got over the Lock, the Diggers taking out the head Solid and making up the Banks.

Mr Greenwood Bridge the Masonry done, and the Wooden Superstructure laid over the same the Cut carried up to the head Solid.

At Mirfield low Hill the Diggers proceeding with the Cut and the same at Ledger Mill,

Mr Gwyn reported that 2 Corn Boats in endeavouring which shou'd be first and also to run down one of our Flatts which was going up before them one of 'em ran foul of an Old Tree laying in Wakefield Dam and staked herself thereupon; but assistance being at hand little more damage ensued than the Wetting about 20 Quarters of the Cargoe of Barley going up for Mr Wilcock. This Tree was attempted to be weigh'd before the Navigation was opened but being deep rooted in the Gravel as well as very large was too much for our Tackle; upon this a Pile was drove down close to the same by way of Guard or Beacon but the same has either by the late Freshes or some Vessel running against it, been drove away; and in Consequence those Keelmen had sufficient Notice thereof by Mr Gwyn but being more attentive to their own Purposes than their Safety occasioned the above accident but as Keels are now upon the River Mr Gwyn proposed and has orders to attempt to weigh the same by means of a Couple of Keels.

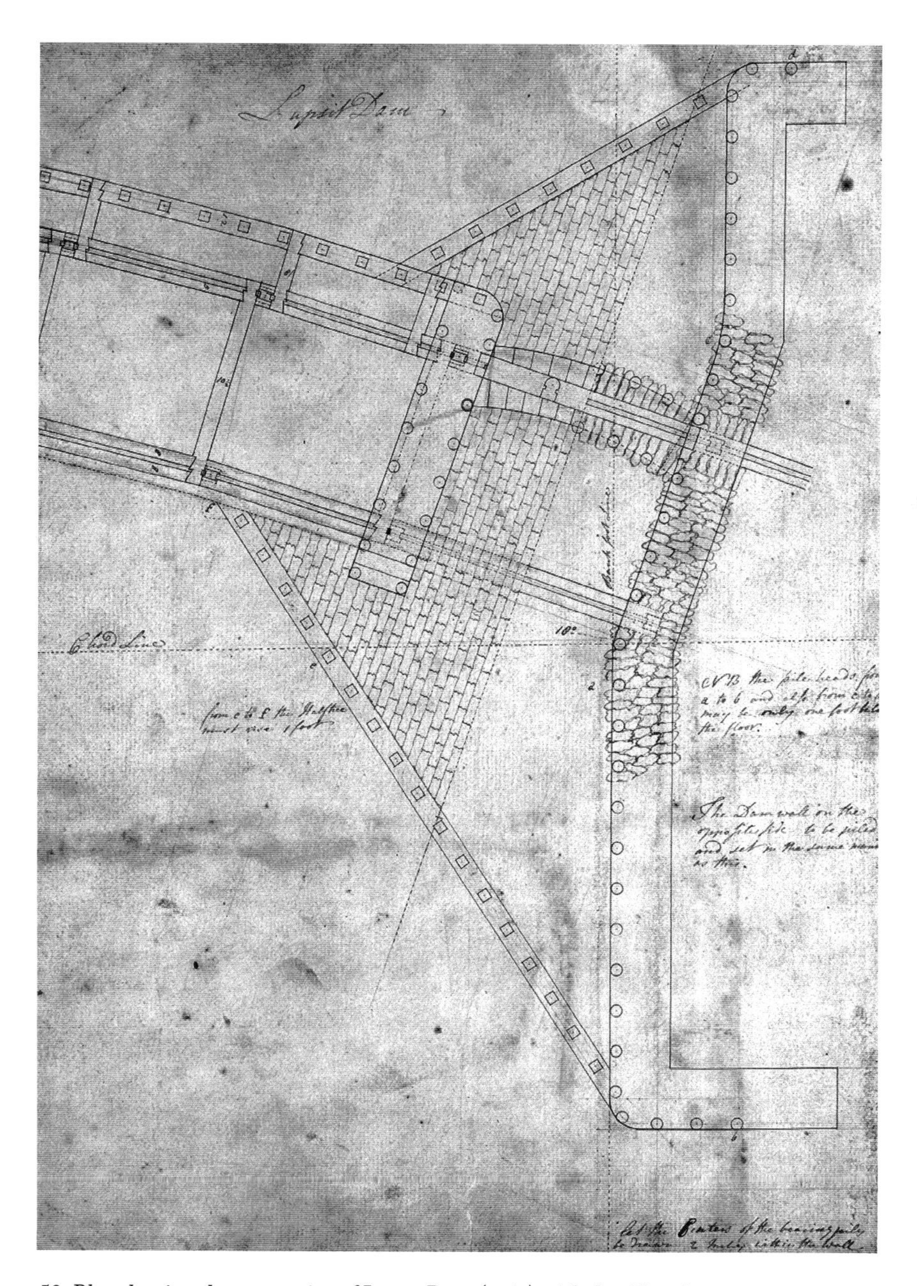

59. Plan showing the construction of Lupset Dam (weir) with the piling shown.

1763

Week 1 March 28th - April 3rd

Monday 28th Tuesday 29th Wen'sday 30th
These Days upon the Works which I found in the following State & Condition:

At Wakefield everything as I left it. About 20 Yards of the South Bank from the Lock Head upward will need raising and strengthening so as to be about a foot higher.

At Thorns the Water in the late Flood had been within a few Inches of the Top of the Bank on the North side from the Lock head to Thorns Bridge and in one place had run over so as to make a small Breach which being observ'd by Mr Gwyn at the time of the Flood was prevented from enlarging; this is repaired as before, but as it appears that the whole Bank is become considerably lower by Settlement it will require raising from the Lock to the Bridge being about 60 or 70 Yards so as to be about Eighteen Inches higher than at Present.

At Lupsit Dam, some of the Dam Boards taken off, a little of the Earth taken from behind the north End Wall and some sods and about a Cart Load of Earth disturbed from a sodded Bank just above the Dam, also some Impression made upon the natural Bank a little below the Dam's Tail just below Lupsit Bridge all on the North side.

At the Tail of Horb[ur]y Mill Pasture Cut a slight Narrow Bridge of Sand thrown across the same but which will be blown out as soon as the Cloughs of the Lock are drawn;

At the Head of Pasture Lock beginning from the North Wing Wall a large Breach extending up the Cut 115 feet and in the deepest part near 8 feet deep below the Dam's height; this Breach extends itself through the Solid from the Cut to the River in Length 315 Feet and being 199 feet broad at the River the whole Area of Ground taken away is 1 Acre & 11 Perches. In consequence of so considerable a Breach a great Proportion of the Rivers Water finding a more direct Passage through the Cut than over the Dam made its way through the same with great Rapidity which of consequence wore away the sides of the Cutt to a considerable Width and formed the Bottom to an inclined Plane from the Rivers Bottom above the head of the Cut to the Rivers Bottom at the Tail of the Breach so that a considerable Part of the Banks of the Cutt had tumbled in after they had withstood the Flood by undermining the Ground whereon they rested.

As it does not appear that this Flood was higher than some the preceding Winter when this Cut was open the breach of the Banks is to be attributed to their being shaken by the Frost before, and being naturally of a loose contexture & this Flood coming thereupon at the same time they were water soaken by the melting of the Snow produced the Effect of first breaking down the artificial Part, and which in all probability wou'd have stopp'd there, had not the surface of the Ground been broke by cutting a Dam in the direction where the Breach is made, for taking off the Water while the Cutt was made in order to save the expence of Pumping. This Drain was well filled up but not having had time to Grass over let the Water within the Surface and which by the long continuance of the first Flood and a succession of several others before the Course of the Water coud be stopped gradually widened itself to its present Dimensions. This Breach happened on the Night between Thursday 17th & Friday 18th of Feb[ruar]y last upon the sudden Melting of one of the largest Snows that has happened in these parts for several Years being as I was informed 20 Inches thick upon the Plain. The Bulk of which having melted off in 2 or 3 Days produced this Flood which yet upwards upon the River was by no means considerable but downwards by the influx of the intermediate Beck was much greater being at Wakefield within some Inches of the Oct[obe]r Flood 1761.

On Tuesday 22 the Waters being considerably abated the Workmen begun to form a Catch Dam near the Head of the Cut by Bringing down and throwing in wet Rubble and Aisler was in Hand at Adingforth Pasture and the work was brought so as to expect a Close on Tuesday 29th but a Fresh coming down and such high Winds ensued that the Flatts coud not work for several Days together so that the work coud not be brought to a final close till Monday following Mar[ch] 7th. During this Time for 2½ Days the whole River went down the Cutt, so that Horbury Mills were totally deprived of Water. The Catch Dam being compleated & the Mills restored to their Water Mr Platts proceeded to throw a Dam of Earth across the Cut below the Catch Dam, so as to put a total Stop to the Water and to be high enough to be Flood proof; and at the same time Mr Gwyn proceeded to drive Piles for forming a temporary Dam to pen the Water while the Earth & Rubble Dams were clear'd out when all the rest were made good. All these Operations appear to have been conducted with the greatest Despatch that the urgency of the case wou'd admit and conformable to directions and Ideas that were interchanged between myself the Surveyors and principle Workmen who have been greatly assisted by W[illia]m Charnley[317] with his Men and Flatt.

The Cut being greatly widened and a great part of the Banks tumbled in as before mentioned in order to procure fresh matter for forming of new Banks it was thought most advisable to widen the Cut till a sufficiency of Matter can be procured, not only to make them up to the height as before but as the same sort of loose Gravel must still make Principal Bulk to make them double in Bulk and Strength to what they were before. Upon my present View thereof I found the greatest part of the Banking on each side the Cut below the Earth Dam performed, the Cut in the widest Place being made 50 Yards and the matter brought forward across the Breach so as to leave only the deepest part of the Channel, the Ground work of which I saw performed with 3 Bays of Plank Piling across the same in the heart thereof, 10 Bays of Pile Dam at the head of the Cut were drove and 8 remained to be done so that by such times as those are clos'd the Breach is expected to be made good so as to proceed to take away the Earth and Catch Dams. One Piece of Service however the Breach has procured the Navigation Viz the Shoal that before laid without the head of the Cutt is now converted into deep Water.

---

317 William Charnley was the undertaker for several locks.

At Washingstone everything in order.

At Horbury Bridge some Sand lodged at the Tail of the Cutt which may be blown out by the Cloughs. The Water has begun to bite upon the Land abreast thereof which may be secured by a little Rubble. The Banks of this Cut which being suspected were raised and strengthened, have stood quite well. At the Breach here nothing has happened worthy of mention save that some of the Rubble that composed the Crown have been conveyed down so as to form a larger Slope & some of the thin Stones that made the Flooring upon the Dams Cloughs had been taken off, but which I found made good.

At Figure 3 everything right and from thence to Thornhill Lock all the little Damage effectually repair'd, but about 100 Yards of the Bank upon the North Side will require raising and strengthening.

Everything made good and in order from Thornhill Lock to Noel Pond and observed the long Cut (though never to be called faulty) yet much more water-tight than when I left it. About 50 Yards of Bank on the north side opposite Sand Mill Dam will require levelling raising and strengthening.

At Noel Pond the Bank from the Warren house Hill to Wilcocks Bridge which was the part suspected to give way if any did was considerably rais'd and strengthened before Christmas so that this standing and the Bank above the Bridge giving way by the Pressure of the Back Water into the Cutt the Stream set directly upon the North Abutment of the Bridge which was piled in front to prevent settlement, but being calculated only for still Water was unable to resist such a Torrent acting immediately upon the Foundation, this produced the Destruction of the North Abutment and in consequence let down the Wooden Superstructure, and this Breach was not only the Cause of this but of every other damage upon the long Cutt. This Bank is now effectually repaired and a Clough hole fixed for draining off the Back Waters at Noel Pond so as to be in no further danger of damage from Floods in that Place, the Bridge was repairing and above water.

At Cut head Dam everything as I left it but needs some addition of Rubble as soon as the Season will permit.

Everything above in order as I left it.

Dewsbury, Mirfield low Mill and Ledger Mill Cutts compleated so far as they can be till the respective Works at the Head thereof are taken in Hand, and Battys Mill Cutt in great forwardness, the Workmen being taken from thence to Horbury Mill Pasture. It will be adviseable further to strengthen the Banks at the Head of Dewsbury and Mirfield low Mill Locks.

On Wen'sday Kirklees Mill Lock Pit was begun. Messrs Wilson & Charnley have both raised a considerable Quantity of Stone for their respective Works.

1763 Week 2 April 4th – 10th

Wen'sday 6th, Thursday 7th, Friday 8th, Saturday 9th
These Days upon the Works and attending the proposed Meeting of Commissioners on Thursday.

All below Pasture Cut as p[er] last. The Breach entirely closed and as high as the Lock Walls, but requires still raising and strengthening.

The Pile Dam clos'd in and the Earth Dam being in a great Measure removed the Calder's Water remaining penned by the Dam of Piles. This Dam is in length near 180 Feet in some Places 9 F[ee]t water against it and the leakage thro[ugh]out the whole not half as much as cou'd be drawn by one hand Screw, and is a Master Piece of Work of its kind.

The Catch Dam of Stones is getting out and everything preparing for opening the Cut. The Calders Water being now within Compass of our Dams Cloughs drew down Lupsit and Washingstone in order to lay bare the Tails of Pasture and Horbury Bridge Cuts; at the tail of Pasture we found a great Quantity of Sludge lodged in the River below the Cuts Tail that had washd down from the Breach which was soon cleared away by the run of the River upon it and it seems that that which is lodged in the Tail of the Cut will soon follow as soon as we are enabled to draw the Cloughs of the Lock.

At the Tail of Horbury Bridge Cut the Cloughs being drawn wash'd out much of the matter but having got a good deal of Gravel therein from the tumbling down of the Natural Bank can not be bottomed without Hands which will be more necessary as the short Water Season last year being over before this Lock was opened cou'd never be thoroughly Bottomed, it is also proposed to cut off a projecting Point of the Land and to slope and stone the foot of the Bank.

Everything from thence to Kirklees as p[er] last, there the Lock Pit almost down to the Water. Set out Cooper Bridge Dam and Lock and the Bridge at Dewsbury Cut Head.

This Week W[illia]m Charnley got up his Flatt over 3 Dams with great Labour & Trouble, it having gone down with himself and all his Hands to assist in making the Catch Dam at the Breach and which was of the greatest use and consequence on that Emergency.

1763 Week 3 April 11th - 17th

Friday 15th, Saturday 16th
These Days out upon the Calder Business being detained the former part of the Week by a Cold and Rheumatic Complaint catch'd by falling into the Calder the Saturday before.

All below Pasture Cutt as before.

The Breach being made up sufficient to stand the Hazard of the Floods while the Workmen are employ'd in still further strengthening thereof. The Earth and Catch Dams being cleared out on Saturday the Water was let in and 2 Bays of Piles from the Pile Dam being drawn Rob[er]t Hartley's Boat went up to Dewsbury loaded with Lime.

On Saturday also was compleated the clearing of the Lock tail at Horbury Bridge.

At Battys Mill the Workmen having worked upon the Lock there all this Week, had got on the Head Sell and raised the Walls several Courses.

At Coopers Bridge the Lock Pit begun on Tuesday last.

At Kirklees the Lock Pit near down.
At Lillands the Lock Pitt begun.

<u>1763 Week 4 April 18th - 24th</u>

Tuesday 19th, Wen'sday 20th, Thursday 21st
These Days upon the Calder, some Hands employ'd at Wakefield in raising & strengthening the Banks near the Lock Head there.

At Pasture Hands employed in strengthening the Bank of the Breach and facing thereof with Stone to prevent the Swell of the Water arising from the Wind from Washing the same.

Mr Wilcocks Bridge made Passable and nearly compleated.

Hands at Work upon the Bridge Pit at Dewsbury Cut Head.

Battys Lock advancing being nearly up to high Water mark.

Coopers Bridge Lock Pit advancing,

Upon levelling for Kirklees Mill Lock Pit I found it 9 Inches too Shallow, the Workmen set on to get out the same.

Lillands Lock Pit advancing.

On Wednesday a Keel went up loaded with Corn and Rape and unloaded at Walkers Bridge for Mr Banks[318], she touched at Wakefield and Washingstone Lock Tails, occasioned by the Sands lodged therein by the Floods, since then Mr Gwyn has been employ'd with Hands in blowing out the same.

---

318 Mr Banks was the owner of Dewsbury New Mill.

1763 Week 5 April 25th – May 1st

Thursday 28th, Friday 29th, Saturday 30th
These Days upon the Calder Works. Thursday Afternoon the Weather begun to be so bad as to put most of the Workmen off work; which continuing all night brought down a Fresh on Friday Morning, which put a Stop to everything below the Calders Water, a second Rain falling on Friday Evening and Night kept everything under Water on Saturday.

The State of the Works as follows.

At Wakefield Cut Tail on the waining away of the Water by the dryness of the Season a large Quantity of Sand appeared to have lodg'd itself above at and below the same. As this Cut Tail had always preserved itself clear from our first working here and even from the original Survey and nothing appearing while the River had a Sufficiency of Water it was not suspected, so that the Bed of Sand was not discovered till the Passage of the first Vessels up the River after the opening of the Breach, and which at first did not appear very material, however the proper means were immediately put in practice in order to blow away the same as mentioned in the Journal of last Week and which so far proved effectual as to make a Channel sufficient for the Passage of Vessels more than once; but as in dry Weather the Cloughs of Kathorp Dam[319] upon the Old Navigation below are drawn every Morning in order to flush up the Boats, this lays the Reach up to Wakefield Lock Tail in a manner dry; So that by the Action of the Water from Wakefield Mills above the Sand lodged above and abreast of the Cuts Tail was driven down into the Navigable Channel, so that what was done one day was undone again the next.

In the dilemma it was resolved by Messrs Smeaton Gwyn & Scott to remove an old Boat that laid at the downstream Point of the Cuts Tail (which had formerly been sunk there for the preservation of the Bank to the upstream Point of the Cut; so as to form part of a Jetty or Wear, and to proceed to compleat the Same with Rubble with all possible expedition in order to prevent the Washing down of the Sands into the Navigable Channel as aforesaid in the execution of which some Steps were taken, but the Calder Freshes have for once conceded to do us a favour for on Sounding the Channel on Saturday Morning the Sand bed was in a great Measure taken away leaving in the shallowest Places above 6 Inches more Water than in the old Navigation Lock and those being of no great extent the Workmen were order'd to proceed to clear the same with the Rake level with the Sell of our own Lock; As this Lock Tail always kept very clear before it is most probable the Sand lodg'd here at this time proceeded from the Breach above and the Body of Sand once got rid of, it is possible it may continue clear, but as we are not sure that the whole of it is come down, and to prevent such Effects for the Future the Removal of the Boat and forming of the Wear is order'd to be proceeded upon as soon as the Freshes will permit.

The Banks of this Cut near the Lock are raised, but some lesser Jobs are still wanted thereat.

---

319 Kaythorp Dam - Kirkthorpe Dam.

The Pasture Bank still further strengthened but not completed.

Greenwoods Cut Head Bridge Pit about 4 ft Calders Water, but filled by the Fresh.

Set out the Cut through Mirfield low Mill Fold and the Bridge Set for Ledger. The Masons proceeding upon Ledger Mill Lock but beat out by the Water.

Battys Mill Lock the Caping got on and all the Masonry done except one of the Head Flue Walls.

At Coopers Bridge the Lock Pit within 2 ft of Bottom but filled.

The Piling of the Bottom of Kirklees Lock in hand but stopp'd by the Fresh.

On Friday The Lock Pit at Brighouse was begun.

Lillands Lock Pit within 2 ft of bottom but also filled.

<u>1763 Week 6 May 2nd - May 8th</u>

Thursday 5th, Friday 6th, Saturday 7th
These Days upon the Works, the Former Part of the Week having proved very rainy, stormy & Cold with constant Freshes little cou'd be done,

At Wakefield the Cuts Tail cleared on Friday and the Flatt & Hands removed to Thorns Lock Tail which tho[ugh] not complainable yet not quite clear.

At Pasture Bank the Hands and Carts there proceeding to form a Road necessary cross the Breach and behind the Bank so as to further strengthen the same; from thence to Dews[bur]y Cut head all as before, the Bridge Pit there got down ready for Piling; the Drainage there remarkably small.

At Mirfield Low Mill Cutt head the Diggers at work upon the Cut head thro[ugh] the Mills fold.

At Ledger the Lock Walls advancing.

At Battys Lock all the Masonry compleated.

At Coopers Bridge the Lock Pit wants about a Foot of its Bottom, one of the new long Hand Screws employ'd here with good Success.

At Kirklees Lock the Sell laid down and the Carpentry advancing; the Screws ordered to be sunk lower and the Lock Pit cleared of Mud left by the Fresh and some Matter not sufficiently cleared to be got out.

Set out Anchor Pit Lock and levelled all the River from thence to Taghole Stream above Lillands.

At Brighouse Lock Pitt some progress made there by the Lillands Hands during the Freshes but now standing.

At Lillands the Lock Pit wants about 18 Inches of the Bottom.

<u>1763 Week 7 May 9th - May 15th</u>

Wednesday 11th, Thursday 12th, Friday 13th, Saturday 14th
These Days attending the Calder Works and at Halifax to attend the proposed meeting of Commissioners. The State of the Works as follows Viz.

At Wakefield all clear Hands employ'd in getting Stone at Agbridge Quarry for forming a Jetty or Wear at the Lock Tail. The Flatt and Hands employ'd in dredging Thorns Lock Tail and compleated the same this Week.

At Pasture Bank righting up and and sodding the same. The Rubble that had formed the Catch Dam at Horbury Bridge brought to the main Body so as to pen the Water over the Dam.

Examined several Pieces of Ground at Dewsbury for a temporary Warehouse and reported the same to the Committee.

Dewsbury Cut Head Bridge Pit the Carpenters Piling it.

Diggers at Work upon Mirfield low Mill Cut head through the Fold.

At Ledger Lock the Walls some part at the Square.

Some Hands at work in Batty's Mill Cutt.

Cooper Bridge Lock Carpenters Piling it.

At Kirklees Mill Lock D[itt]o and compleated it, this Work ready for the Masons.

Anchorpit Lock Pitt begun on Wednesday.

Brighouse d[itt]o standing still.

Lillands Lock Carpenters piling it.

<u>1763 Week 8 May 16th - 22nd</u>

Friday 20th, Saturday 21st
These Days upon the Calder Works.

At Mill Pasture a Flat employ'd bringing down Rubble from Addingforth Quarry Pasture for securing the South Bank; some hands righting up the same.

At Washingstone the other Flatt employ'd in dredging the Locks Tail into which some Matter had been deposited by the Winters Floods for want of a Sufficient Jetty or Wear from the Upstream Point.

All from thence to Dewsbury Cut head as p[er] last; at the Bridge Pitt there the Carpenters compleated and Masonry begun.

At Mirfield low Mill Cut head the Digging compleated through the fold, and for the Foundation of the Turnbridge, and delivered a Design for the same.

Ledger Lock the Masonry completed.

At Coopers Bridge Lock the Carpenters Work in the foundation far advanced.

At Kirklees Lock the Masons at work since Monday;

Anchor Pit Lock Pit advancing below Calders level.

Brighouse Lock Pit standing still.

Lillands the Carpentry in the Foundation far advanced.

<u>1763 Week 9 May 23rd - 29th</u>

Friday 27th, Saturday 28th
These Days upon the Calder Works.

At Wakefield Lock Tail removing the old Boat from the downstream Point in order to form a Wear from the upstream Point.

At Washingstone Lock the dredging compleated on Friday and a Rubble Wear from the Upstream Point of the Cuts Tail in a great Measure formed.

At Horbury Bridge a Wear of like Nature formed there, and Hands employ'd in Strengthening the Crown of the Rubble Dam at the Breach.

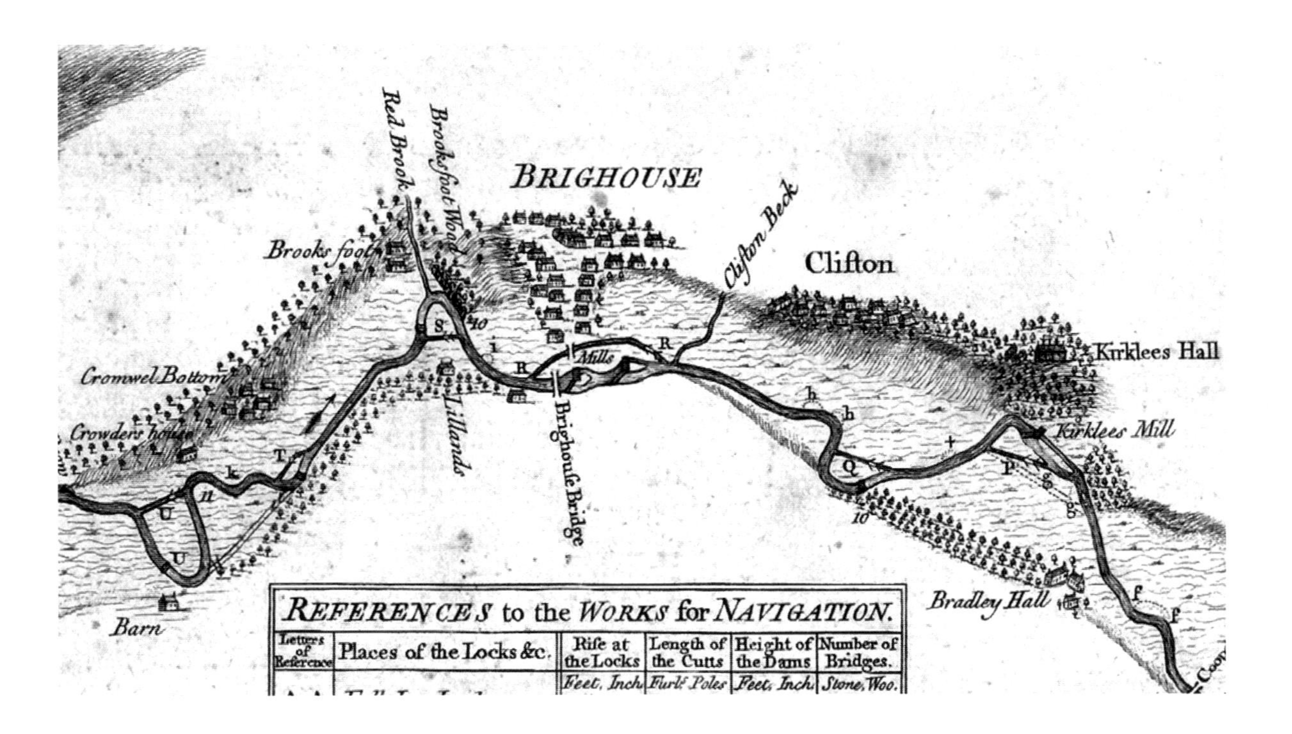

60. Brighouse, Kirklees and Lillands from John Smeaton's plan of 1757/58

At Dewsbury Cut head the Bridge above Water.

At Mirfield Cut head on levelling the Bottom found it too Shallow by 8 Inches.

At Ledger all as p[er] last. Mr Platts bro[ugh]t Barrows & Planks to Cap the Bridge Pitt the former Part of this Week, but was prevented by Mr Banks threatening him that if he put Spade into the Ground till he was satisfy'd for damages he woud sue him.

At Batty Mill the Hands proceeding with the Cutt.

At Coopers Bridge Lock the Piles all drove and rough set but the Sell wants planking and Threshold laying. At Coopers Bridge Dam the Principal Row of Gage Piles drove across the River and Carpenters proceeding therewith.

Kirklees Lock Walls above Water and setting the Bottom.

Anchor Pit Lock Pit was bottomed on Friday, and Carpenters begun Piling on Saturday.

At Brighouse the Hands resumed the Work at the Lock Pit on Saturday.

At Lillands the Masons begun at the head of the Lock on Monday last but the Carpenters did not compleat the lower Sell till Saturday.

<u>1763 Week 10 30th May - 5th June</u>

Thursday 1st, Friday 2nd, Saturday 3rd
These Days attending the Calder Works and at Halifax to attend the proposed Meeting of the Commissioners.

At Wakefield Lock Tail Hands at work in forming a Wear and clearing the Old Boat which is wrecked up so fast that it is obliged to be pulled in pieces and removed Piece Meal at such times as the drawing of Kathorpe Dam Cloughs will permit.

At Thorns Hands employ'd in raising the Banks of the Cutt from the Lock Head to Thorns Bridge which are become too low by settlem[en]t as formerly mentioned.

At Pasture Bank the South Bank fronted with Stone and proceeding with the Sodding about the same.

At Horbury Bridge Hands employ'd in making up the Crown of the Rubble Dam so as to be 18 Inches above the Crown of the main Dam.

At Dewsb[ur]y Cutt head Bridge the Hands proceeding with the same above Water.

At Mirfield low Mill Cut Head the Hands staking out the Ground 8 Inches deeper as mention'd in my last; This Ground being too high was occasioned by an Error in laying the Upper Sell of Mirfield Lock and which by this means I discovered; The Sell was laid according to my own order; and the Error was not in taking the level having done that twice over which nearly agreed, but as I apprehend in casting up the Same. However I found it within the Compass of being rectify'd without disturbing the Sell or impairing its Strength; a considerable Allowance having been made in the structure of all those Sells to correct an accidental Error; which has not been necessary in any former Case.

At Batty Bridge Cutt the Hands proceeding therewith

At Coopers Bridge Lock the Carpentry compleated and the Masonry considerably advanced since Monday; At the Dam the Carpenters proceeding with the Pile Work.

Kirklees Lock being got above Water the Water was let in and all Hands taken off to Cooper's Lock.

At Anchor pit the Carpenters proceeding with the Piling.

At Brighouse the Hands digging the Lock Pit which turns out a very hard Clay and Gravel.

At Lillands Lock the Masonry advancing.

1763 Week 11 6th - 12th June

Thursday 8th, Friday 9th, Saturday 10th
These Days upon the Works.

At Wakefield the Tail Wear compleated but the Old Boat not quite removed. Some Hands preparing Stone for raising the Breast Work in the Lock Head a Course higher.

Hands at Thorns proceeding with the Banks.

At Mill Pasture the Spade Work of the Banking compleated.

At Horbury Bridge the Hands proceeding to form an Upstream Slope from the raised Crown of the rubble Dam.

Dewsbury Cutt head Bridge standing still for want of Stuff.

Upper Sell at Mirfield rectify'd by lowering the same 8 Inches leaving a 7 Inch Threshold remaining. At the Cut head the Piling begun for the wharf Walling thro[ugh]the Mill fold and for the Turnbridge;

At Ledger Mill the Bridge Pitt begun; Hands proceeding with Batty's Cut;

The greatest Part of Coopers Lock above low Water Digging out for the Cloughs at the Dam, making temporary Dams; and proceeding with the Piling.

Kirklees as p[e]r last;

Anchor Pit Carpentry far advanced and Masons rough pitching under the Walls.

Brighouse Lockpit proceeding as per last;

Greater Part of Lillands Lock above Water.

1763 Week 12 13th - 19th June

Friday 17th, Saturday 18th
These Days upon the Calder works:

At Wakefield Lock the Old Boat removed, the Wear compleated & the Breast Work raised a Course higher.

Hands at Thorns at work raising the Banks.

At Horbury Bridge the Rubble Dam compleated.

At Long Cut head Dam removing Gravel from the Lock Hill in order to back up the Dam.

Watergate Lock Carpenters employ'd in hanging the Tail Gates.

At Dewsbury Cut head Bridge standing still as p[e]r last for want of Stone.

At Mirfield the Piling for the Turn Bridge and Wharf Walling thro[ugh] the Mill Fold compleated, the Trunk for the House Drain laid down and Masonry begun.

At Ledger Cutt head the Bridge Pit near down,

At Batty Cut Hands proceeding with the same & with the Bridge Pit.

At Coopers Bridge the Lock being above water the Masons begun upon the North land Wall of the Dam for the Cloughs; and the Carpenters proceeding with the Piling;

Kirklees Lock being above Water standing as p[e]r last;

At Anchor pit the Piling of the Lock Pit compleated ready for the Masons;

Brighouse Lock Pit advancing under Calders level.

Lillands above Water and the Water let in, Carpenters driving the Gage Piles at Lillands Dam

1763 Week 13 June 20th - 26th

Tuesday 21st to Saturday 25th
These Days upon the Calder Works, at the Union Club and attending the Annual Meeting;

Tuesday and Wednesday attended S[i]r Geo[rge] Savile who survey'd all the Works from Wakefield Lock to Elland

The State of the Works on Saturday as follows.

At Thorns hands employ'd in raising the Banks as p[er] last from thence to Long Cut head doing nothing, at long Cut head backing up the Dam as p[er] last,

Everything on Dewsbury Cutt as [per] last.

At Mirfield Cut head the Masonry advancing.

At Ledger Cutt head the Bridge Pit got down.

At Batty's Cut the Bridge Pit got down and all hands off;

At Coopers Bridge Dam Masons at work upon the Conduit for the Cloughs and Carpenters at work upon the Piling and framing.

Hands begun at Lyon Royd stream to dredge that Shoal, from thence to Brighouse all as p[er] last,

At Brighouse the Lock Pit compleated on Saturday.

Lillands Lock standing still as p[er] last the Hands employed at Mirfield Cut head. The Carpenters proceeding with the Piling of the Dam.

1763 Week 14 June 27th - July 3rd

Friday July 1st, Saturday 2nd
These Days upon the Calder works.

The Banks at Thorns being raised to their proper Height and the Stone Breasting paved at Top. The Hands were departed and nothing doing between Wakefield and Dewsbury; This Week the Sheds from Wakefield Yard were taken down and carried up to Dewsbury in order to be erected for temporary Warehouses in Mr Wilcocks Orchards as p[er] order of the Commissioners.

At Dewsbury Cut all as p[er] last;

At Mirfield Cut head the Masonry up to the square and working at the Breast Walling; the Carpenters beginning to hang the Gates at the Lock.

At Ledger Cut head hands getting the Water out of the Bridge Pit, in order to begin Piling the same;

at Battys Cut the Bridge Pit piled and the Masonry in hand;

At Cooper Dam the Cloughs Conduit compleated & the river turn'd thro[ugh] the same, several Bays of the Dam the Masonry upon a close, and the Carpenters proceeding with the rest;

At Lyon Royd stream the Shoal there in great Measure compleated and clearing the Tail of Kirklees Lock and Reach below.

Hands at work in clearing the Shoal at Thistley Stream in the Reach between Anchor Pit and Brighouse in the general Plan call'd Willow and Lee Streams;

Brighouse Bridge Pit standing for want of Carpenters.

Lilland Lock as p[er] last; The Carpenters at the Dam here gone down to Coopers Bridge, the Labourers digging out for the Cloughs Conduit.

1763 Week 15 July 4th - 10th

Thursday 7th, Friday 8th, Saturday 9th
These Days upon the Calder Works and at Halifax to attend the Meeting of Commissioners.

At Long Cut Head Dam Hands employ'd in Backing the same.

At Wilcocks Orchards Hands getting up the Sheds;

At Dewsbury Cutt head Masons proceeding with the Bridge and the same got ready for the Carpentry on Saturday.

At Mirfield Lock Hands employ'd in hanging the Gates; the Bridge etc at the Cutt head as p[er] last;

At Ledger Cut the Bridge piled and the Masons proceeding with the same,

At Batty's Cutt Carpenters laying down a Tunnell for discharging Water from the adjacent Lands: everything else on this Cut as p[e]r last;

At Coopers Bridge Hands proceeding with the Cutt; At the Dam the Carpentry all compleated the Body of the Dam expected to be clos'd in on Saturday night and the South End Wall to be rais'd above Dam's height.

At Lyon Royd Stream and Kirklees Lock Tail the Dredging all compleated.

At Anchor pit Hands begun to dig out the Ground for the Dams Cloughs.

At Thistley Stream the dredging expected to be compleated on Saturday night;

At Brighouse the Lock Pit standing still for want of Carpenters;

At Lillands Lock nothing doing the Hands employ'd on Ledger Bridge.

At Lillands Dam Hands digging out for the Conduit Cloughs and Carpenters proceeding with the Piling.

1763 Week 16 July 11th - 17th

Friday 15th, Saturday 16th
These Days upon the Calder Works;

At Horbury Bridge Hands employed in forming a Road across the Breach in order that the Horbury People may get off their Crops as heretofore from their Pasture on the South side of the River.

At Dewsbury Hands employ'd upon the temporary Warehouses and levelling the Ground and making a Road.

At Watergate Locks Hands employ'd in dredging the Lock Tail and leading the Matter of the Lock hill in order to strengthen the Banks at the Lock head. At the head of the Cutt the Timber Bridge got on and the Masons employd in getting up the Wing Walls.

At Mirfield Lock Hands employ'd in hanging the Gates; at the Cut head the Work standing still for want of the Carpentry of the Swivel Bridge.

At Ledger Bridge the Work up to the square but standing still for want of Centers.

At Battys Cut the Masonry of the Bridge advancing, and the Wood Tunnel laid for the Drainage.

At Coopers Bridge the proper Hands proceeding with the Lock, Cut and Dam. On Saturday Morning being a Fresh the Cloughs cou'd not vent the Water, so that the Dam ran over for the first time;

Anchorpit Carpenters proceeding with the Piling for the Dam.

At Brighouse Lock all as p[e]r last;

At Lillands Dam the Carpentry advanced so that the Masons were expected to begin on Monday.

<u>1763 Week 17 July 18th - 24th</u>

Saturday 23rd

This Day upon the Calder Works but little Business done this Week on acc[oun]t of the Rains and Floods; being a severe Rain on Tuesday a considerable Flood came down the Calder on Wednesday the largest that has been in Summertime since the Works were begun; None of the finish'd Works have rec[eive]d prejudice, nor any of the unfinish'd that is of the least consequence.

The Backing of the Dam at long Cut head with Gravel & stone compleated;

Masons building up a Wharf Wall with dry Rubble at Dewsbury Warehouse. The Sheds there completed so far as they can be for want of Deals[320], which have been wrote for to Hull sometime since but no answer rec[eive]d;

The Hands proceeding with the Crane in Wakefield Yard.

At Dewsbury Cut head Hands employ'd in levelling the Ground and forming a Road over the Cut head Bridge.

At Mirfield Lock the Gates hung all else as before.

At Ledger Bridge as p[er] last. Hands taking out the Lock head Solid.

At Batty's Cut the Bridge as p[e]r last, Masons employd in making Drains leading to and from the Tunnel;

At Cooper's Lock the Mason at work and part of the Caping on; Coopers Dam in the same situation as the last Week, any further than the northland Wall being left open and unbacked. The Water during the Flood worked behind it, and it is surprizing the River had not made its Way into the Cut and there form'd a new Course. This Event I pointed out to Mr Wilson the preceding Week, and expressly ordered the Wall to be made up immediately, and as I frequently inculcated during the Course of building this Dam the great Damage that might happen from attempting a close before the Land Walls were compleated in case of Floods; I hope this narrow Escape from a similar damage to that at Horbury Bridge will teach Mr Wilson the necessity of this Doctrine;

At Anchorpit Dam no Damage except the temporary Dams and Engine carried away by the Flood. From thence to Lillands all as p[er] last.

320 Deals - fir or pine boards (Ref: Oxford Dictionary)

At Lillands Dam no Damage except to the temporary Dams; but those were repaired and Carpenters at work.

Mr Smeaton having set out the Works given the necessary Plans & Directions for proceeding during his Absence took leave of the Works this Day in order to set out for Scotland next Morning.[321]

1763 Weeks 18-22 July 25th- August 28th No entries. *Smeaton in Scotland.*

1763 Week 23 August 29th - September 4th

Tuesday Aug 30th, Wednesday 31st, Thurs Sept 1st, Friday 2nd
Arrived at Austhrop from Scotland on Monday Evening. These days at Halifax and attending the Calder Works which I found as follows.

During my Absence of 3 Weeks there has been the most remarkable rainy Season that has been known for several Years[322] almost all the Kingdom over scarce ever 2 Fair Days together, and scarce a Week but one or more Floods upon the Calder which have greatly retarded the Progress of the Works that might have been expected had the Weather proved favourable. However on viewing all the Navigation Part from Wakefield to Dewsbury I did not find any Damage but everything in good order. The most remarkable Accident that I have been informed of was the Breaking loose of the new Flatt from her Moorings upon the sudden rise of a Flood in the night which being loaded with Timber chiefly belonging to the Contractors but some to the Commissioners which driving over 2 Dams a considerable Part of both was lost; particularly several material Parts of the Crane for the temporary Warehouse at Dewsbury which have been obliged to be made new yet I found the Warehouse there and Crane nearly ready for use all except the Flooring with loose Deals which have never yet been able to be procured from Hull.

At Watergate Lock found the Water just let in upon it and the Cut made Navigable; the Banks appear very light but a considerable Leakage thro[ugh] the Walls of Mr Greenwoods Bridge therefore ordered Back Drains to be compleated immediately to carry off the Water till the runs at the Bridge can be stopp'd, a small Leakage also from behind the North Wall of the Lock owing to the Filling the Lock Wall not being made so compact as it shou'd have been, the upper Gates being also more leaky than they may be, ordered them to be taken out and cased. Ordered also a Pair of Flood Gates to be made for the Flood Gate Bridge at the head of the Cut to be made to prevent damage to the Banks etc by Floods.

321 This visit resulted in plans for bridges at Perth and Coldstream, and the first plans for the Forth and Clyde Canal. (Ref: Skempton: John Smeaton, Trevor Turner and A. W. Skempton. p 18)

322 1763 was noted as a very wet summer in England and Wales, during which the Thames flooded (Ref: http://www.pascalbonenfant.com/18c/geography/weather.html)

No Water in the Cellar of Mr Greenwoods House on Friday Morning but some got into the Cellar of the Mill House.

The hawling Way and Bridge over Ravens Brook made passable and the hawling Gates putting up from Dewsbury to Mirfield Lock.

At that Lock the Gates hung and the Turning Bridge at the Mill Fold made passable and the Diggers getting out the Solid at the head ready for Walling the Cutt above the Bridge. The Rains have driven down much Sand into the lower Part of the Cutt which needs deepening.

At Ledger Lock the Solid at Lock Head got out and the Bridge at the Head of the Cutt ready for removing the rubbish in order to make the same passable.

At Batty's Mill the Masonry of the Bridge not quite up to the square the Tunnel compleated and all the Solids out except the Cuthead and that for a temporary Road at the Bridge.

At Coopers Lock the Masonry compleated except the Head Flue Wall; At Cooper's Dam all in good order except that the Floods had taken down a part of the Partition Wall between the Cloughs and Body of the Dam owing to its being left uncap'd. The Cut completed all except the head Solid.

At Kirklees the Masonry of the Lock completed, except the Flue Walls, the Diggers proceeding with the Cut as fast as they can till the Crops are off.

At Anchor Pitt the Mason closing in the Arch for the Dams Cloughs, the Water getting out of the Lock Pit in order to begin the Masonry,

the dredging Work at Lyon Royd & Thistley Streams in good Order.

Brighouse Lock not touched since my last; The Diggers going on with the Cut;

at Lilland the Lock as p[e]r last, but the Dam the Timberwork nearly compleated, the Conduit compleated with the partition Wall and the 2 Dams End Walls rais'd up to the height of the Dams Crown & 2 Bays of the Body of the Dam set out at each End.

<u>1763 Week 24 and 25 September 5th - 18th</u>

Thursday 8th, Friday 9th, Saturday 10th, Sunday 11th, Monday 12th, Wednesday 14th, Thursday 15th
These Days attending the Calder Works,

All below Dewsbury as p[e]r last, the temporary Warehouses & Crane at Dews[bur]y ready for use; and compleated so far as they can be till Deals are got for compleating the Floors, ab[ou]t ? of the Floor of the principal Warehouse being laid.

At Watergate the back Drains making; the Runs stopping at Mr Greenwood's Bridge, the Banks strengthening near the Bridge and the Carpenters forming the Flood Gates for the head of the Cut.

At Mirfield the Hands being at work clearing the Cut, and staking out the Solid at the head thereof, a Flood which came down on Friday the 9th filled the same, but the Gates being hung no damage ensued; the decline of the Ground near the Lock not having afforded a sufficiency of Matter to make the Banks there proof against the very high Floods, and being of a very sandy Nature, the Carts were at work leading the Matter of the Lock Hill in order to strengthen the same in consequence of a former order; also order'd the breasting (formerly order'd) at the Lock head to be forthwith apply'd.

Everything at Ledger & Batty's Works as p[er] last.

At Coopers Bridge Dam, the Conduit Wall made up well caped and cramped.

The Diggers at Work on Kirklees Cutt;

At Anchor Pit the Dams Cloughs being compleated and part of the Conduit Wall caped, the Flood on Friday made its Way thro[ugh] the same, and at the Lock the Masonry having been advanced about 2 Courses high and the Floor set, the Flood made its way into the Lock pit, and made a small Breach occasioned by the Workmen's endeavouring to keep out the Calder till it rose too high; no other damage down here; on Wednesday the 14(th) the Masons building up the north End Wall of the Dam to its proper height, and the Carpenters preparing to resume their Work at the same.

At Brighouse the Diggers at work upon the Cut.

At Lillands on Thursday the 8th found the Body of the Dam clos'd in, the South End Wall rais'd to its proper height [see next week notes], but the North End Wall rais'd no higher that the Dams Crown; to this I remonstrated having the previous Week ordered both the End Walls to be rais'd to their proper height, back'd in with Earth and the Ground secured before the Dam was clos'd in; however it having been exceeding fine Weather for some Days there was no appearance of immediate danger; yet in the night between the 8th & 9th the Weather changed and a great deal of Rain fell on Friday Morning and bro[ugh]t down a Flood, which abated on Friday Afternoon without damage; accordingly I went down the Works to Dewsbury recommending it to the special Care of Mr Scott & Wm Charnley to have a watchful Eye over this Dam; who report that on Friday Eve'ning everything was as it had been all the Day and therefore was thought safe; but a fresh Rain falling that Night which occasioned a fresh rise of water, on Saturday Morning at break of Day, the Water had made its way behind the Dams North End Wall and at 10 o'Clock when I got thither it had taken away ab[ou]t a Rood of Ground and was working down the Land very fast;

Saturday prov'd very rainy bad Weather, yet a Number of Carts having been procured and a temporary Stable having been built of dry Rubble for the Gin Horses near the Place by the help here of a Quantity of Rubble was speedily procured, which put a Stop to the incroachment of the Water, and before night the Ground was so far secured that tho[ugh] the Water increased on Saturday Ev'ning no further Advances were made;

The Carts and Men were kept going on Sunday; and in a few Days the Breach might have been stoppd and everything secured, but being Harvest and the Roads from the Quarry's being cut so deeply by working in the Rain that it was impossible to keep the Carts to work, however with such as cou'd be procured the Work was carried on, and on Wednesday the 14th Matters were bro[ugh]t to a good Posture of defence, the Weather being then good and the River greatly abated, I left them with order to make a Rubble Bank from the South End Wall and Main Land and to make everything secure there before they attempted to further fill up the Breach.

This is the second Accident of the kind that has happened besides a very narrow escape at Coopers Bridge notwithstanding my repeated Orders in every Dam that has been built, to get up the Land Walls and secure the Ground at the Ends before a Close of the main Body of the Dam is attempted; but not to lay blame on Particulars we have been continually press'd with such a Quantity of work going on at a time that 'tis with the greatest difficulty that anything can be got compleated in due time and as it shou'd be.

<u>1763 Week 26 September 19th - 25th</u>

W 21st, T 22nd, F 23rd
These Days attending the Meeting of Commissioners at Halifax and on the Calder Works.

On Sunday Ev'ning fell a remarkable downfall of Rain which on Monday Morning produced the highest Flood this Year, and continuing rainy all that day & Tuesday the Waters continued high till Wednesday Morning but without any damage to the Works.

At Watergate Cut the Labourers at Work in making a drain on the north side and the Carpenters at work upon the Flood Gates. Upon the Rain & Flood and not before, some Water came into Mr Greenwoods Cellar, but as the Water was abated therein on Friday & these Cellars have always been subject to receive Water in time of Floods, it still remains a doubt whether they are liable to be affected by the Water of the Cut or not, the Cellar of the Old House being much nearer the Cut has had Water ever since the Cut was opened which increased upon the Flood & likewise on Friday had abated.

At Mirfield Lock the Carpenters employ'd in fitting up the lower Gates they having been put into Place but not finished. The last Flood's Water when at the height was within 2 ft of the Top of the Locks Wall so that the Breasting already ordered appears the more necessary, as the highest Floods have been known almost 3 ft higher.

At Ledger, Battys & Cooper's Bridge the Works all as p[e]r last.

At Kirklees the Diggers going on with the Cut;

At Anchorpit the Diggers having repaired the Breach made into the Lockpit by the Floods on Friday the 9th Inst & got the Water out on Sunday in order to be ready for the Masons on Monday morning, a fresh Breach was made on Sunday Night & the Lockpit refilled but without further damage. The Dam & every thing here as p[er] last;

At Brighouse the Diggers going on with the Cutt.

At Lillands the Rubble Bank before ordered to be constructed between the north End Wall and the main Channel (in my last Journal by Mistake called the South End) across the low Ground being made & the Breach being clos'd on Friday the 16th every thing here stood the last Flood without further damage & on Friday 23rd the Breach was made up almost to Dam's height Viz. as high as it ought to be till the Body of the Dam & Clough Walls are compleated. The Securing of this Work in this manner so early greatly depended on 3 Carts supplyed by Messrs Platts & Burgon[323] & one kept by Wm Charnley, the greatest Part of the Country Carts having given over on account of the badness of the Road from the Quarrys.

1763 Week 27 September 26th - October 2nd

Sept 29th and 30th
These Days upon the Calder Works;

At Watergate Cutt the Banks in part strengthened and the Carpenters hanging the Floodgates.

At Mirfield Cutt strengthening the Banks and the Masons about raising the Breastings at the head of the Lock. The Dredgers clearing the Tail; this Week several Vessels up at Mirfield low Mill and one d[elivere]d part of a Cargo of Lime at the Tail of Ledger Lock, but it will be impossible to clear this Breach effectually this Winter on account that it cannot be expected that the Cloughs of Mirfield Lock will be able to command the Water. This hindrance has been entirely owing to the Carpenters who being engaged with the Foundation of the New Work were unable to get the Gates of that Lock hung and the Turning Bridge in Place, so as to be in a capacity of turning the Water at the Lock till after the Rainy Season commenced; however we hope to maintain the necessary depth of Water whilst the Dams continue to run over and after that to dredge it down to its proper depth.

I found by the level of the Water left by the Flood in Ledger Cutt, that the upper Sell of that Lock lay 4ft? under the Level of Ledger Dam but being informed that the Millers frequently

323 Jeremiah Platts was a witness to William Burgon's marriage to Mary Hepworth at Horbury on 2nd July 1763.

draw down above one Foot order'd the Sell to be cut down 6 Inches which can easily be done without Prejudice while the Water is off the Lock and before the Gates are hung. Nothing done here since the last.

At Battys Lock the Upper Gate hung & the Bridge Walls rais'd up to the Square ready for the Carpentry.

All at Coopers Bridge as before.

At Kirklees Cut the Diggers going on with the same.

At Anchor Pit the Weather having continued fair from Friday to Sunday last, the Drainers according to order repaired the Breach and got the Water out of the Lock Pitt, and the Masons were at work thereupon and also the Masons at work upon the South End Wall of the Dam and the Carpenters at the Piling.

At Brighouse the Diggers upon the Cutt;

At Lillands the Masons at work making up and Caping the Dams End Wall next the Breach.

<u>1763 Week 28 October 3rd - 9th</u>

Oct 6th, 7th, 8th
These Days at Halifax to attend the Comm[issione]rs and upon the Calder Works.

This Week the Carpenters compleated the Flood Gates at the head of Watergate or Dewsbury Cutt, the Carts strengthening the Banks and a beginning made for a drain in order to take off the Water from a part of a Close contiguous to the Cut; whose Surface lays 1ft? below the level of the Water in the Cutt and which having no way to get off is rendered poachy and wet.

At Mirfield Cut the Breasting and Banks at Lock Head made up but some low places above still want strengthening. The Dredgers having made way for the Vessels at the Tail of the Lock for the present are now working at the head of the same.

At Ledger Cutt all as p[er] last.

At Batty's Cutt the Masons making up the Breasting at the Lock head the Carpenters at work at the Bridge being got on the Carts at work leading matter from the Lock hill to form a Road over the same.

All at Coopers Bridge as before.

At Kirklees the Diggers going on with the Cut,

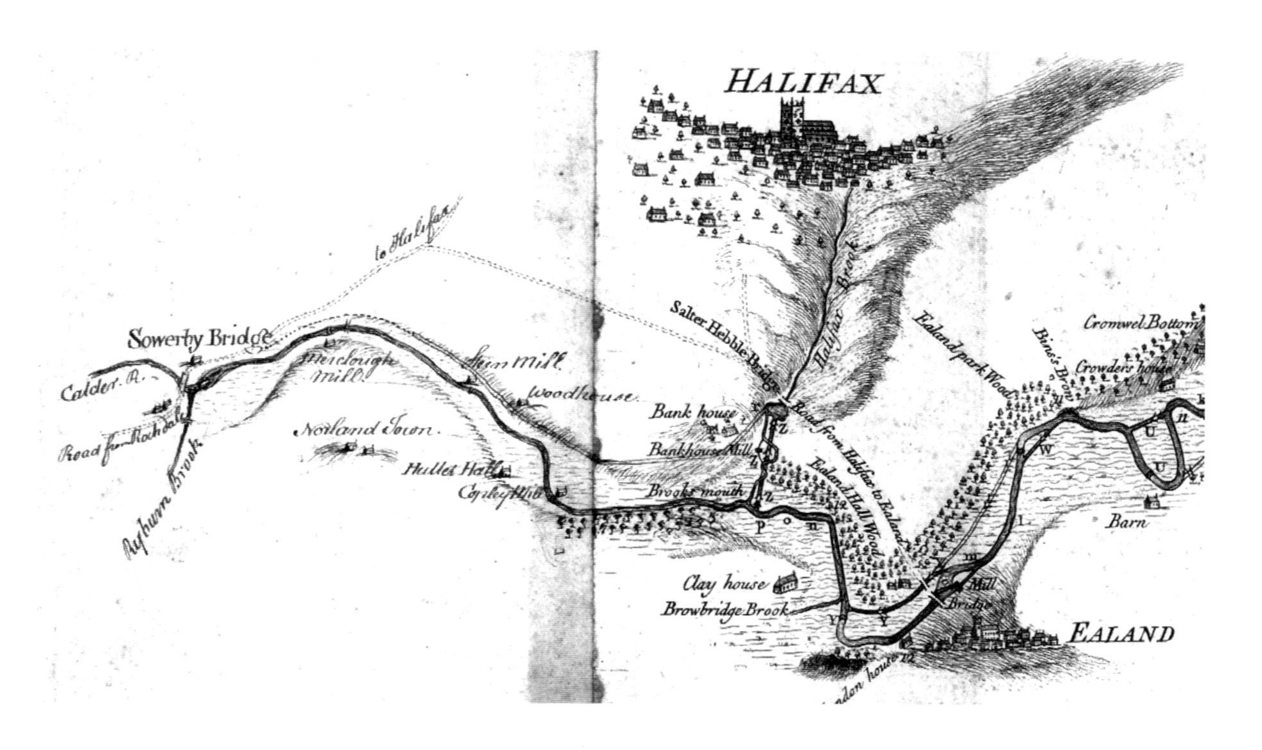

61. Halifax and Sowerby Bridge from John Smeaton's Plan of 1757/58 with the later addition from Brooksmouth to Sowerby Bridge.

At Anchor Pit an heavy Rain having fallen in the Night between the 1st & 2nd Inst the Lock pit was again filled and the Earth Dam broke down; and this not having been repaired this Week the Masons were at work upon one of the side Walls which had been got above Water, ordered the Masons to proceed therewith and the South End Wall of the Dam till the breach cou'd be repaired and the Water got out, but in the Night between Friday & Saturday an heavy Rain falling in the West on Saturday came down a Bank full of Water.

At Brighouse Lockpit the Water got out & Carpenters begun piling the same this Week but all drove out by the Water.

At Lillands the South End Wall the Caping got on but wanted Cramping the Masons proceeding with the Body of the Dam which on Friday wanted 3 Bays of a Close.

1763 Week 29 October 10th - 16th

Octr Th 13th, F 14th, Sat 15th
These Days upon the Calder Works.

In consequence of the heavy Rains that fell on Friday and Saturday last the Water continued high till Monday and on Thursday Morning Mr Gwyn came to acquaint me with a Misfortune that had happened to Lupsit Dam which he had discovered the Day before upon the Wain off of the Flood Water; his acc[oun]t was as follows that the setting between the Skirt and Crown was dropped down into the Body of the Dam for 5 or 6 Bays somewhat nearer the South End than the Middle, in consequence whereof the Water fell perpendicularly from the Crown which was standing entire, into those Bays where the Setting had failed.

That on Saturday last in going up the River he passed by the Dam before the Flood came down and in appearance all was right. When I got there on Thursday the Water had made a thorough Breach having entirely carried away one Bay and greatly damaged two More, three others being standing in the Manner Mr Gwyn had described; from these circumstances it appears Evident that this Misfortune was entirely occasioned by the Floods having moved a part of the Rubble deposited at the Skirt and had formed a Pool which had loosened the feet of the Skirt Piling; and thereby letting out the filling of the Dam the setting wou'd drop and the Water acting perpendicularly wou'd in like manner pool the Feet of the Skirt Heart Piling; after which the dependencies being gone the Bays wou'd necessarily give way at once. The Skirt of this Dam was entirely rubbled before the last Winter, and after the Spring Floods were over, the Flatts were employed carrying down Rubble to rectify what was amiss, this made me rely that everything was safe before the late rainy Season, but since my Letter to Mr Simpson on this occasion, I find John Smith[324] & Co who were employ'd on this Service had declared that they had discovered while working there a hole at the Skirt of the Dam pooled near 8 ft deep which they did not fill up for want of Stuff & for want of particular orders thereupon, that they mention'd the same to me in the presence of Messrs Gwyn & Scott but that I made no answer nor took no Notice; this is certain that neither Mr Gwyn nor Mr Scott nor myself remembers the least little of such information; and it is plaind (that) it must be delivered in some Way unheard or unattended to, otherwise such a circumstance wou'd not have passed unnoticed.

The Season being far advanced and the Weather too precarious to think of a thoro[ugh] Repair this Year I ordered the Breach to be blocked up and secured as well as possible with Rubble. The Principal difficulty consists in this that the Water of the Breach being in a great measure drawn off thro[ugh] the Breach the Flatts cannot get down from the Quarrys with even half loading; however all the 3 were got to work on Friday with some additional help.

All from hence to Ledger Cut as p[er] last. The upper Sell of this Lock rectifyed as before ordered; the Carpenters at work putting together Gates and the Bridge being made passable the Diggers getting out the Head Solid.

At Batty's Cut Hands strengthening the Banks near the Lock Head;

---

324 In 1767 John Smith became the resident engineer under Smeaton for the River Ure Navigation and canal to Ripon, with Joshua Wilson as contractor for locks.

From hence to Anchor pit all as p[e]r last. At Anchor pit Lock the Dam being repaired and the Water got out on Wednesday, the Masons had got both the side Walls above Water, had set the upper Sell and were getting the head of the Lock out of Water; The Diggers clearing part of the Lock in order to compleat the setting and clearing the Foundations for the last flue Walls. The Dam as [per] last.

At Brighouse Lock the Water had been got out this Week and the Carpenters at work piling the same but the Horses failing the Water put a Stop to present proceedings;

At Lillands the Roads from the Quarrys having become so bad that nobody work their Teams the Undertaker got up his Flatt by land Carriage on to Lillands Reach, and having opened a new Quarry for Rubble by the Riverside, the Rubble wanted for compleating and securing this Work will now be got on very moderate terms. On Saturday the Dam having suffered nothing from the late Floods was within a Day of being closed.

<u>1763 Week 30 October 17th - 23rd</u>

October T 18th, W 19th, Th 20th, Sun 23rd
These Days upon the Calder Works.

The 3 Flatts having Plyed hard with stone on Saturday Sunday & Monday last on Tuesday the Breach at Lupsit was grounded and blocked up to such an height as to admit of their Passing with full Boats and Thursday the Navigation was again restored; several Vessels having passed up & down the River and the Flatts continued at work strengthening the Breach so that from the time that Orders were given for its repair to the time that Boats had free Passage was exactly a Week.

But the repair of the Misfortune of the Dam has been productive of another which tho[ugh] not of any great consequence will yet occasion some fresh trouble, Lupset Reach being drawn off by the Breach very little Water was left upon the Floor of Pasture Lock and in this case there being 12 feet head of Water upon the Tail Gates the drawing of the Cloughs for passing of the Flatts and flushing them out of the Lock has disordered the setting of the Stern Apron and some of the Setters were drove out of place, tis true the drawing of the Cloughs with so great head and so little Tail Water woud rush out with uncommon violence but yet had the Setting been well performed it is not easy to conceive how this shoud have happened. Yesterday (viz Sunday 23) I went over to examine the State therefore to give Orders for its amendment.

All from hence to Ledger Lock as before. At Ledger the Carpenters getting in the Fore Gates and at work upon the Stern Gates, Diggers getting out the head Solid.

At Battys Lock strengthening the Banks from thence to Anchor pit nothing doing.

At Anchor pit all the Walls of the Lock out of Water the upper Sell set the Tail flue Walls raising and the Diggers getting out the Tail Solid.

At Brighouse the Carpenters piling the Lock Pit

At Lillands the Body of the Dam clos'd and getting Rubble to back up and Skirt the Dam and to make up the Breach and back up the Dam's End Wall.

1763 Week 31 October 24th - 30th

25th, 26th, 27th, 28th
These Days upon the Calder Works and attending the proposed meeting of the Commissioners at Halifax.

At Lupsit Hands & Flatts at work strengthening the Body of a matter in the Breach and lining the Skirt of the Dam the whole Mass of Rubble that before lined the Skirt having subsided from End to End and an old tree of Black Oak appeared cast up upon the Gravel Bed below upon which the Skirt Piles had probably been driven the displacing whereof might be the immediate Cause of the failure in the part where it happened.

Hands employed in baring Stone & Quarrying Rubble at Adingforth & Mill Bank Quarries in order to strengthen and secure this and the other Works; which tho[ugh] upon examination they don't appear to have suffered by the late Flood yet in point of Prudence they ought to be farther guarded.

At Cut head Dam 2 or 3 Flatt loads of Rubble deposited there in order to secure the same.

At Mirfield Cutt Hands at work compleating the Banks.

At Ledger Lock Carpenters hanging the low Gates the Diggers getting out the Solid at the Tail, Bottoming & clearing the Cutt and getting out the Solid at the Head.

At Battys Cutt the Bridge made passable and Diggers getting out the Solids left for temporary Roads.

At Coopers Bridge Dam the Ground being pool'd below the Skirt tho[ugh] not much more than after the first floods yet sufficient to hold a Body of Rubble therefore order'd the Rubble laying for that Purpose at the South End to be laid in immediately and a further Quantity to be got in order to secure the same compleatly.

At Anchor pit Lock everything below Water compleated & the Water let in the Walls almost all up to the Square, the Dam as p[er] last.

At Brighouse the Carpenters piling the Lock and the Masons getting up the Walls at the head of the Lock.

At Lillands the Dam backed and skirted and the Breach made up so as to pen the Water over the Dam but wants strengthening and also some Rubble to secure the tail of the South End Wall as well as the land on the Northside all which were ordered to be done forthwith.

1763 Week 32 October 31st - November 6th

Oct 31st, Nov 1st, 2nd, 3rd
These Days upon the Calder Works

On Monday Mr Gwyn sent a Message that that morning a very sudden Flood had come down the River which he was afraid had done a further prejudice to the Dam at Lupsit; accordingly I went over and in the Afternoon the Water being then considerably abated did not find that anything material had happened; the Main difference being the moving of some of the Rubble from the Top of the Breach.

On Monday Evening & night fell a good deal more rain and next morning there was a full Water upon the Dam but had the Satisfaction to observe there was no material alteration hence I proceeded up the River and found everything from thence to Ledger Lock as p[er] last, here I found the Water upon the Lock without having done any damage. Mr Platt having duly executed the Order I gave him last Week Viz that the Solid at the Cutts head being taken down while it wou'd but just pen the Calders Water if a Flood shou'd suddenly come down and break in, it might endanger the blowing up of the Bridge and therefore on the approach of a Flood to let in the Water gradually which was done on Sunday night the Gates of the Lock being then hung the Head Gates breasted and the Banks made up, the Tail Solid got out as far as it cou'd be by digging but some of the Solid at the head left in.

At Battys Lock the Carpenters upon the Gates, from hence to Anchor pit all as before.

At Anchor pit part of the Caping on; the rest nearly up to the Square, the Dam as before.

At Brighouse Lock the greatest part of the Chamfer'd Course on, at the head some Courses above & part of the Floor set, when the whole when the whole [sic] was interrupted by the Flood on Sunday night.

At Lillands Dam nothing there had stirr'd altho[ugh] the Water had run over the Land in to the Lock, in consequence whereof ordered a Rubble Bank to cross the Neck of Land from the Rubble Dam to the Bank of the Cutt to be strengthened and the Aisler at the Lock Head to be disposed bank fashion to prevent the Floods from running over the Solid into the Lock.

Finding all safe above and having given the necessary Orders I returned on Wednesday to Lupsit, the Water being somewhat abated found the Dam in still the same condition and on visiting it again on Thursday found no apparent change, but as the Waters from the violent showers that accompanied the Squalls of Wind for the 3 preceeding Days were not likely speedily to abate so as to come to a full Inspection I departed the Work with Orders to Mr Gwyn to continue to set down all the Stone he cou'd to the place in order to be ready to throw in to strengthen such parts as appear most to have suffered, as soon as the Water woud permit, and on returning on Sunday found the Water greatly abated and Mr Gwyn with all the hands throwing Stone to strengthen the Whole which I had the Satisfaction to see had suffered nothing material and much less than cou'd have been expected.

1763 Week 33 November 7th - 13th

10th, 11th, 12th
These Days upon the Calder Works.

At Lupsit the Breach made up to the proper height which suffered no derangement from a considerable Fresh which happened on Tuesday, the Hands proceeding to strengthen the Backing and Skirting of the Dam,

the Diggers proceeding as ordered to strengthen a part of the Bank at Horbury Bridge Cutt which had somewhat settled so as to become full low as also some places upon the long Cutt.

Some Hands at Mirfield Mill Fold putting the same to rights as we found it and the Masons making up & compleating the Walls and the Sides of the Cutt.

At Ledger Mr Platt having made up the Bank at the head of the Cutt in order to bottom the head Solid attempted to draw off the Water, but found that the Bank was sapped at the bottom, he therefore was obliged to let the Water in again; the Cutt head must therefore be piled across in order to enable him to bottom the Solid, which was ordered.

At Battys Lock hanging the Gates; At Coopers Bridge Lock the Dam putting in Rubble for securing the Skirt

At Kirklees Lock Masons letting in the Iron Work & Diggers upon the Cutt

At Anchor pit all as p[er] last

At Brighouse Diggers at work upon the Cutt.

At Lillands the Rubble Bank compleated as ordered last Week but on Sunday last there being a fresh in the River, in lowering down the Flatt to deliver its Cargo upon the Breach one of the hands by mistake letting go the Rope by which it was fastened the Boat drove into the Cloughs and immediately sank with its Cargo of Stone. On Saturday Wm Charnley having lightened her with Draggs was endeavouring to weigh her but this appeared hardly feasible till it was somewhat further Lightened she did not appear to have done any damage to the Works; but this Work not being as yet in a State of Security order'd Mr Scott to get up one of our own Flatts to compleat the same in case it shou'd be found impracticable to weigh her or that she shou'd be found unserviceable when weighed.

BIBLIOGRAPHY

Primary Sources

The Superintendent's Journal is part of a large collection: Calder and Hebble Navigation Company 1756-1936, held at the National Archives: RAIL 815. For background to the Journal itself, other related documents in the same collection and covering the same period have been extensively referenced from the microfilm at Calderdale Archives: MIC2/1 Committee minutes letters and subscriptions 1756-1758; MIC2/2 Commissioners' Minutes 1758-1858; MIC2/4 letters and reports 1758-1797; and MIC2/22 Journal 1760-1765 (which includes accounts).

Smeaton's detailed drawings of locks, weirs and other structures are available at the Royal Society in London: Smeaton Drawings, JS/6 Volume VI Canals and River navigation 1750-1792

The Journals of the House of Commons and the House of Lords provide details of the progress of the Bills through Parliament including the petitions and evidence given to the Committees. The West Yorkshire History Centre in Wakefield holds copies.

The Cavendish Nevile Papers are available through the Yorkshire Archaeological and Historical Society, now at the Brotherton Library, Leeds: YAS MD 335/3/10. Calderdale Archives holds a number of letters, subscription certificates and other related papers. Kirklees Archives and Dewsbury Library hold a number of maps and deeds related to Thornhill. The Rockingham Collection is at Sheffield Archives.

Newspapers were accessed through the online British Newspaper Archive, directly by subscription, through Find My Past, or through some local libraries which offer free access to their members. Other local newspapers are available on microfilm at Wakefield Local Studies Library although no specific Wakefield paper exists for this period.

Books

Skempton, A.W. (ed), *John Smeaton FRS*, Thomas Telford Limited, London 1981

Hadfield, Charles, *The Canals of Yorkshire and North East England*, 2 Vols, David and Charles, Newton Abbott, 1972

Hadfield, Charles, *The Canal Age*, David and Charles, Newton Abbott, 1968

Smith, Peter L. *The Aire and Calder Navigation*, Wakefield Historical Publications, Wakefield, 1987

Taylor, Mike, *The Calder and Hebble Navigation*, Tempus Publishing, Stroud, 2002

Chrimes, Michael M, *The Civil Engineering of Canals and Railways before 1850*, Routledge, 1998

Hargreaves, John A, *Halifax*, Edinburgh University Press/Carnegie Publishing, 1999

Inland Waterways Association, *Guide to the Calder and Hebble Navigation and the Huddersfield Broad Canal*, undated.

Taylor, Mike, *The Canal & River Sections of the Aire & Calder Navigation*, Wharncliffe Books, Barnsley 2003

Parry, Keith, *Trans-Pennine Heritage: Hills, People and Transport*, David & Charles, Newton Abbot, London, 1981

Transactions of the Halifax Antiquarian Society 1901- present. Available in bound volumes at Calderdale Central Library, Halifax

Duckham, B. *John Smeaton and the Calder & Hebble Navigation*, Railway and Canal Historical Society Journal Vol 9 1964.

Holmes, John, *A Short Narrative of the Genius Life and Works of the late Mr John Smeaton*, London 1793

Watson, Garth, *The Smeatonians, the Society of Civil Engineers*, London, 1989

Skempton, Chrimes, *A Biographical Dictionary of Civil Engineers in Great Britain and Ireland,* Vol 1, London, 2002

Redmonds, George, *Names and History*, Hambledon and London, 2004

Smeaton, John, *John Smeaton's Diary of his Journey to the Low Countries*,1755, Leamington Spa,1938, (printed for the Newcomen Society)

Vallancey, Charles, *A Treatise on Inland Navigation, or, the art of Making Rivers Navigable, of Making Canals in all Sorts of Soils, and of Constructing Locks and Sluices*, 1763, Dublin, (reprinted by Gale Eighteenth Century Collections Online)

Online books

Online at Archive.org: https://archive.org/index.php

Crossley, E.W. (ed) *The Monumental and other Inscriptions in Halifax Parish Church*, John Whitehead & Son, Leeds, 1909.

Royds, Sir Clement Molyneux, *The Pedigree of the Family of Royds*, Mitchell Hughes and Clarke, London 1910

Stansfeld, John, *History of the Family of Stansfeld of Stansfield in the Parish of Halifax*, Goodall and Suddick, Leeds, 1885

Online at Google Books: https://books.google.co.uk

Nicholson, Peter, *The Builder and Workman's New Director*, Edinburgh, 1845

Rees, Abraham, *The Cyclopedia; or Universal Dictionary of Arts, Sciences and Literature* Vol 38, 1819

Millington, John, *Elements of Civil Engineering*, Dobson, Philadelphia, 1839

Mahan, D.H. *An Elementary Course of Civil Engineering*, Wiley & Putnam, New York, 1838

Online websites

Malcom Bull's Calderdale Companion
http://www.calderdalecompanion.co.uk

From Weaver to Web, Online visual archive of Calderdale's History
https://www.calderdale.gov.uk/wtw/index.html

Engineering Timelines
http://www.engineering-timelines.com/timelines.as

Limestone Quarrying and Burning in Leicestershire, Barrow-on-Soar Heritage Group: http://www.busca.org.uk/heritage/articles/old-industries-of-barrow/limestone-quarrying-burning-in-leicestershire.html

Barwick in Elmet Historical Society
www.barwickinelmethistoricalsociety.com

Watchet Conservation Society
http://www.watchetconservationsociety.co.uk/newsletter.html

LIST OF ILLUSTRATIONS

11. Contemporary drawing of John Smeaton's Eddystone Lighthouse. Wikimedia Commons.
12. Excerpt from Leeds Intelligencer 13 September 1757. Newspaper image © The British Library Board. All rights reserved. With thanks to The British Newspaper Archive: (www.britishnewspaperarchive.co.uk).
13. Excerpt from John Smeaton's Report from *Minutes of Committee to consider proper measures to obtain an Act of Parliament for making River Calder navigable from Wakefield to Elland and so on to Halifax.* The National Archives, RAIL: 815/1.
14. Panel listing shoals and streams from John Smeaton's Plan of the River Calder 1757/58. ©Royal Society, John Smeaton Archive: JS 6/9.
15. *Reasons for Extending the Navigation of the River Calder from Wakefield to Halifax.* West Yorkshire Archive Service, Bradford. Ref: SpSt/13/2/2.
16. The White Hart in Wakefield. From a map of 1823, surveyed by J. Walker.
17. John Smeaton's Plan of the River Calder 1757/58. ©Royal Society, John Smeaton Archive: JS/6/9.
18. Cartouche from John Smeaton's Plan of the River Calder 1757/58. ©Royal Society, John Smeaton Archive: JS/6/9.

The Bill in Parliament

19. *Plan of that part of the River Calder that lies between Sowerby Bridge and Halifax.* John Eyes, 11 January 1758. 'Weaver to Web' online, document ID: 100442.
20. Panel listing mills, locks and dams from John Smeaton's Plan of the River Calder 1757/58. ©Royal Society, John Smeaton Archive: JS/6/9.
21. Title page of the River Calder Navigation Act 1757. Wakefield Local Studies Library. Ref: Calder and Hebble Navigation, Rules and Acts of Parliament, Wakefield 368.4 Y.

From the Passing of the Act to the start of the Journal

22. Subscription of Japhet Lister, Merchant of Halifax of £10, 19 June 1760. West Yorkshire Archive Service, Calderdale. Ref: HAS 153 668/99. Reproduced by courtesy of Halifax Antiquarian Society.
23. Photograph of Austhorpe Lodge. Leeds Library and Information Service, Leodis. Subject ID: 8688.
24. Drawing of William Jessop. Wikimedia Commons.

Quarries

all maps from Ordnance Survey of the 1850s.

25. Heath
26. Newmillerdam
27. Altofts

Wood Supplies

Gathering Materials

Buying Land

Workers and working arrangements

46. Entry in Account book for 18th May 1760. (from microfilm),WYJS/CA MIC2/22. By kind permission of the National Archives.
47. Excerpt from the Manchester Mercury 23rd February 1762. Newspaper image © The British Library Board. All rights reserved. With thanks to The British Newspaper Archive: (www.britishnewspaperarchive.co.uk).
48. Excerpt from the Manchester Mercury 14th December 1762. Newspaper image © The British Library Board. All rights reserved. With thanks to The British Newspaper Archive: (www.britishnewspaperarchive.co.uk).

After the Journal

49. Title page of the River Calder Navigation Act 1769, 9 Geo 11 c 71. Wakefield Local Studies Library: Calder and Hebble Navigation Rules and Acts of Parliament Wakefield 368.4 Y.
50. Excerpt from the Leeds Intelligencer, 9th July 1771. Newspaper image © The British Library Board. All rights reserved. With thanks to The British Newspaper Archive: (www.britishnewspaperarchive.co.uk).

Transcription of Smeaton's Journal

51. Title page of the Superintendent's Journal 1760. The National Archives: RAIL 815/16 Superintendent's Journal 1760-1763.
52. Excerpt showing Wakefield and Lupset from John Smeaton's Plan of the River Calder 1757/58. ©Royal Society, John Smeaton Archive: JS/6/9.
53. Excerpt showing Horbury from John Smeaton's Plan of the River Calder 1757/58. ©Royal Society, John Smeaton Archive: JS/6/9.
54. General View of a Calder Lock ©Royal Society, John Smeaton Archive: JS 6/18.
55. Excerpt showing Dewsbury and Thornhill from John Smeaton's Plan of the River Calder 1757/58. ©Royal Society, John Smeaton Archive: JS/6/9.
56. Plan of Thornes Lock ©Royal Society, John Smeaton Archive: JS 6/26.
57. Plan of Horbury Bridge Lock, the lower gate sill ©Royal Society, John Smeaton Archive: JS 6/34.
58. Excerpt showing Mirfield and Cooper Bridge from John Smeaton's Plan of the River Calder 1757/58. ©Royal Society, John Smeaton Archive: JS/6/9.
59. Plan of Lupset Dam ©Royal Society, John Smeaton Archive: JS 6/45.
60. Excerpt showing Brighouse and Elland from John Smeaton's Plan of the River Calder 1757/58. ©Royal Society, John Smeaton Archive: JS/6/9.
61. Excerpt showing Halifax and Sowerby Bridge from John Smeaton's Plan of the River Calder 1757/58. ©Royal Society, John Smeaton Archive: JS/6/9.
62. John Smeaton's Plan of the River Calder, surveyed in 1757, engraved in 1758. ©Royal Society, John Smeaton Archive: JS/6/9.

INDEX

JS refers to John Smeaton. *Italics* indicate images, n a footnote. **Bold** page ranges reflect repeated and/or occasional discussion of a topic in the transcription of the journal. A page number may reflect the main discussion of a topic also mentioned many times in passing in the transcription.